Applied Science Review™

# Genetics

Applied Science Review™

# Genetics

Michael Benner, PhD
Assistant Professor
Department of Biology
Rider University
Lawrenceville, New Jersey

Springhouse Corporation
Springhouse, Pennsylvania

## Staff

**EXECUTIVE DIRECTOR, EDITORIAL**
Stanley Loeb

**SENIOR PUBLISHER, TRADE AND TEXTBOOKS**
Minnie B. Rose, RN, BSN, MEd

**ART DIRECTOR**
John Hubbard

**ACQUISITIONS EDITOR**
Maryann Foley

**EDITORS**
Diane Labus, David Moreau, Marguerite Kelly

**COPY EDITORS**
Diane M. Armento, Pamela Wingrod

**DESIGNERS**
Stephanie Peters (associate art director), Matie Patterson (senior designer)

**ILLUSTRATORS**
Jackie Facciolo, Jean Gardner, John Gist, Judy Newhouse, Amy Smith, Stellarvisions

**MANUFACTURING**
Deborah Meiris (director), Anna Brindisi, Kate Davis, T.A. Landis

**EDITORIAL ASSISTANTS**
Caroline Lemoine, Louise Quinn, Betsy K. Snyder

**Cover:** *DNA replication inside a nucleus. Scott Thorn Barrows.*

©1995 by Springhouse Corporation, 1111 Bethlehem Pike, P.O. Box 908, Springhouse, PA 19477-0908. All rights reserved. Reproduction in whole or part by any means whatsoever without written permission of the publisher is prohibited by law. Authorization to photocopy items for personal use, or the internal or personal use of specific clients, is granted by Springhouse Corporation for users registered with the Copyright Clearance Center (CCC) Transactional Reporting Service, provided the base fee of $00.00 per copy plus $.75 per page is paid directly to CCC, 27 Congress St., Salem, MA 01970. For those organizations that have been granted a license by CCC, a separate system of payment has been arranged. The fee code for users of the Transactional Reporting Service is 0874346762/95 $00.00 + $.75.
Printed in the United States of America.
ASR11-010794

Library of Congress Cataloging-in-Publication Data
Benner, Michael
Genetics / Michael Benner.
p. cm. — (Applied science review)
Includes bibliographical references and index.
1. Genetics—Outlines, syllabi, etc. I. Title II. Series
QH440.2.B46 1995
575.1 — dc20 94-13692
ISBN 0-87434-676-2 CIP

## Contents

## Advisory Board

Leonard V. Crowley, MD
Pathologist
Riverside Medical Center
Minneapolis;
Visiting Professor
College of St. Catherine, St. Mary's Campus
Minneapolis;
Adjunct Professor
Lakewood Community College
White Bear Lake, Minn.;
Clinical Assistant Professor of Laboratory Medicine and Pathology
University of Minnesota Medical School
Minneapolis

David Garrison, PhD
Associate Professor of Physical Therapy
College of Allied Health
University of Oklahoma Health Sciences Center
Oklahoma City

Charlotte A. Johnston, PhD, RRA
Chairman, Department of Health Information Management
School of Allied Health Sciences
Medical College of Georgia
Augusta

Mary Jean Rutherford, MEd, MT(ASCP)SC
Program Director
Medical Technology and Medical Technicians—AS Programs;
Assistant Professor in Medical Technology
Arkansas State University
College of Nursing and Health Professions
State University

Jay W. Wilborn, CLS, MEd
Director, MLT-AD Program
Garland County Community College
Hot Springs, Ark.

Kenneth Zwolski, RN, MS, MA, EdD
Associate Professor
College of New Rochelle
School of Nursing
New Rochelle, N.Y.

## Reviewers

DuWayne C. Englert, PhD
Professor of Zoology
Biological Sciences Program
Southern Illinois University
Carbondale, Ill.

Paul Goldstein, PhD
Professor of Genetics
Department of Biological Sciences
University of Texas
El Paso

## Acknowledgments and Dedication

I thank the reviewers for patiently wading through my manuscript. I especially thank the faculty and students at Rider University and Philadelphia College of Pharmacy and Science for their encouragement, insight, and feedback.

To Elisa

## Preface

This book is one in a series designed to help students learn and study scientific concepts and essential information covered in core science subjects. Each book offers a comprehensive overview of a scientific subject as taught at the college or university level and features numerous illustrations and charts to enhance learning and studying. Each chapter includes a list of objectives, a detailed outline covering a course topic, and assorted study activities. A glossary appears at the end of each book; terms that appear in the glossary are highlighted throughout the book in boldface italic type.

*Genetics* provides conceptual and factual information on the various topics covered in most introductory genetics courses and textbooks and focuses on helping students to understand:

- the principles of Mendelian analysis and advanced Mendelian analysis
- eukaryotic chromosomal structure
- the chromosome theory of inheritance
- the structure and replication of genetic material
- the processes involved with linkage, recombination, and gene mapping
- genomes and gene expression in eukaryotes and prokaryotes
- chromosomal mutation
- the nature of transposable elements, retroviruses, and oncogenes
- principles of non-Mendelian inheritance
- population and quantitative genetics.

# 1

# Introduction to Genetics

## Objectives

After studying this chapter, the reader should be able to:

- Discuss the relevance of genetics to basic scientific research, medicine, agriculture, and biotechnology.
- Explain the difference between heredity and variation.
- Distinguish between a genotype and phenotype.
- Give an example of how the environment can affect the ultimate expression of traits.

## I. Scope and Relevance of Genetics

### A. General information

1. ***Genetics,*** a term coined by English biologist William Bateson in 1906, is the study of ***heredity*** and ***variation***
   a. Heredity is the biological similarity between parents and their offspring
   b. Variation is the biological difference between parents and their offspring
2. Genetics includes the study of the mechanisms of inheritance and the net effect of these mechanisms on the functioning of the organism and the species
3. The fields of cytology (the study of cells) and molecular biology (the study of molecular processes) are intimately related to genetics
4. Genetics plays a central role in the study of biology
   a. Genetics embraces all organisms, including viruses, in all kingdoms
   b. Genetics is studied at the molecular, cellular, organismal, and population levels
5. Genetics also plays a key role in human affairs
   a. Agriculture, food science, and medicine depend on genetic research to a great extent
   b. The emerging role of molecular genetic engineering (which is nothing short of phenomenal) is forcing individuals and governments to evaluate the ethical, legal, and sociological ramifications of new technologies
   c. The importance of genetics to society commonly is underscored by the awarding of a Nobel Prize for significant accomplishments in the field of genetics
   d. Nobel laureates in genetics include Thomas Cech (United States) and Sidney Altman (Canada) for their discovery of enzymatic ribonucleic acid (RNA); Susumu Tonegawa (Japan) for research in immunogenetics; Walter Gilbert (United States) and Frederick Sanger (Britain) for techniques of de-

oxyribonucleic acid (DNA) sequencing; and François Jacob (France), André Lwoff (France), and Jacques Monod (France) for research in enzyme biosynthesis

**B. Genetics and basic science**

1. Genetic dissection is a powerful tool used to elucidate life processes
2. Variability in a particular trait can clarify the control of that trait
   a. By studying abnormal traits or abnormal mechanisms of inheritance, we can better understand normal processes
   b. For example, the study of human metabolic abnormalities and deficiencies allows researchers to piece together the chain of events that results in normal metabolic functioning
3. Geneticists often trace the inheritance patterns of abnormal traits to study the mechanisms of heredity rather than the abnormal trait itself; in this way, traits serve as genetic markers

**C. Genetics and medicine**

1. The understanding of the genetic flaws responsible for many chronic and life-threatening diseases opens the door to understanding the biochemical basis of these disorders
2. After the biochemical basis of a disease has been determined, appropriate therapies can be designed
3. Discovering the genetic basis of a disorder also facilitates the screening of additional individuals who may have the disease or are asymptomatic carriers
4. Although direct genetic therapy for hereditary defects is in its infancy, it promises to become a powerful tool for curing genetic disorders, such as cystic fibrosis and muscular dystrophy

**D. Genetics and agriculture**

1. Manipulating plant and animal populations to produce increasingly desirable products efficiently requires a firm grasp of the patterns of inheritance
2. Breeders have long recognized the existence of natural genetic variability in domestic and wild populations; this variability applies to pest resistance, nutritional quality, growth rate, grain yield, and fat content of meat
3. Controlled mating of selected individuals results in the incorporation of desirable characteristics into a single population
4. The design and execution of breeding programs differs widely from one trait to another and from species to species, depending upon the genetic basis of the manipulated traits and the natural breeding behavior of the species
5. Advances in molecular genetic engineering provide additional opportunities for altering plants and animals, thus improving the availability and quality of food

**E. Genetics and biotechnology**

1. In the early 1970s, scientists learned how to splice together the hereditary material of different individuals and organisms
2. Numerous commercial products now are produced by bacterial cells that carry genetic instructions which are foreign to these cells
   a. Exogenous human insulin is derived from genetically engineered *Escherichia coli* cells; this process circumvents the need to isolate insulin from animal sources

(1) *E. coli,* a bacterial inhabitant of the intestinal tract, is a prokaryotic organism that has many uses in basic scientific research and biotechnology

(2) *E. coli* is easily cultured in large volumes of liquid media, thereby making possible the isolation of substantial amounts of metabolic products

b. Additional compounds produced by bacterial cells include growth hormones, vitamins, amino acids, and interferon (a chemical messenger that helps cells resist viral infection)

## II. Variation and Heredity

### A. General information

1. Experimental genetics depends largely upon observation of traits for which discrete, nonoverlapping variants exist
   a. Hundreds of variants, such as curly wings, scarlet eyes, and black bodies, have been studied in the common fruit fly *(Drosophila melanogaster)*
   b. Each variant of this type is readily distinguishable from normal fruit flies
2. In nature, however, it is difficult to categorize much of the variation that occurs
   a. Instead of facilitating the placement of individuals in distinct categories, variations in such traits as size, color, and behavior result in a broad spectrum of appearances that commonly resembles a normal distribution
   b. A more complex, statistical approach is required for illustrating inheritance patterns in this type of variation

### B. Lessons learned from the domestication of plants and animals

1. Our interest in heredity and variation originated with the domestication of plants and animals, which began approximately 10,000 years ago
2. Heritable traits were recognized and manipulated during the breeding process
   a. From the study of detailed Egyptian tomb carvings, scientists have concluded that the first animal species to be genetically manipulated were sheep, goats, camels, dogs, and oxen
   b. Other Egyptian stone carvings indicate that date palms were among the first plants to be pollinated artificially; hundreds of different varieties were produced by the observation and selection of genetic variants
3. Observation of heredity and variation among domesticated species shaped our early views of genetics and probably led to the observation of heritability in humans
   a. The early Greeks were particularly interested in human heredity
      (1) The theory of pangenesis postulated that small particles from all parts of the body accumulated in semen, where they were passed on to the next generation
      (2) Some Greek natural historians, including Aristotle, believed that traits acquired from experience or accident were passed on to offspring
   b. Medieval scholars generally accepted the theory of preformation, in which gametes were thought to contain a completely formed but miniature human being

### C. Gregor Johann Mendel

1. The modern view of genetics was conceived first by Gregor Mendel, an Austrian monk, scientist, and mathematician, who studied at the University of Vienna in the early 1850s
2. After leaving the university, Mendel taught science and undertook breeding experiments with the garden pea at a monastery in Brünn, Austria
3. In the paper "Experiments in Plant Hybridization," Mendel discussed many of his conclusions about the principles of heredity; although this research was published in 1865, it was not until the turn of the 20th century that his theories were widely understood and accepted
4. Hugo De Vries (Holland), Carl Correns (Germany), and Eric von Tschermak-Seysenegg (Austria) independently rediscovered and verified Mendel's principles in 1900

## III. Role of the Environment

### A. General information

1. The development of an organism is the result of its genetic makeup and the environment in which that makeup is expressed (or shown)
   a. Although a corn plant may be genetically programmed to achieve great height, it will not grow to its full potential without a suitable growing environment
   b. Even genetically identical twins develop into unique individuals on the basis of the particular life experiences of each twin
2. Therefore, the development of an organism is viewed as an interaction between genetic programming and the environment

### B. Genotype vs. phenotype

1. The genetic makeup of an organism is called the ***genotype***
2. The net result of an organism's genotype and the environmental impact on that organism is called the ***phenotype;*** thus, the phenotype can be explained as the appearance and functioning of an organism
3. Although genotypes are fixed entities for any particular cell or organism, the phenotype changes according to the organism's environment and individual history of development

### C. Norm of reaction

1. An individual genotype may express a variety of phenotypes in different environments
2. The range of phenotypes expressed by any one genotype is called the ***norm of reaction***
3. Graphs commonly are used to present norms of reaction
   a. The horizontal axis represents the range of environments in which the genotype was tested
   b. The vertical axis represents the range of phenotypes expressed as a result of genotype and environmental interaction
4. The norms of reaction may intersect for different genotypes, thereby indicating that different genotypes exhibit the same phenotype in certain environments

## Study Activities

1. Discuss three ways in which the field of genetics affects our daily lives.
2. Define the following terms: heredity, variation, genotype, and phenotype.
3. Draw a graph that illustrates a hypothetical norm of reaction for an experimental organism of your choice.

# 2

# Mendelian Analysis

## Objectives

After studying this chapter, the reader should be able to:

- Describe Mendel's experimental methods and terminology.
- Illustrate the principle of segregation by diagramming a monohybrid cross.
- Explain the relevance of a testcross.
- Illustrate the principle of independent assortment by diagramming a dihybrid cross.
- Apply the sum rule and product rule to problems of probability.
- Interpret human pedigrees and deduce the associated genotypes.

## I. Mendel's Garden

### A. General information

1. Mendel manipulated pea populations by controlling pollinations
2. Mendel's initial populations were true breeding, meaning that no variation existed within populations
3. The cross-pollination of individuals from phenotypically different populations resulted in the production of ***hybrid*** plants
4. Subsequent crossing of hybrids resulted in characteristic patterns of inheritance in ensuing generations
5. Besides noting the resulting plant phenotypes, Mendel noted the ratios at which those phenotypes occurred

### B. Generation terminology

1. The initial generation from which individuals are selected for cross-hybridization is termed the ***parental (P) generation***
2. The population resulting from the hybridization of parental individuals is called the ***$F_1$ generation*** or first filial generation
3. The population resulting from hybridization of $F_1$ individuals is called the ***$F_2$ generation*** or second filial generation
4. The enumeration of generations goes on indefinitely

# II. Monohybrid Crosses

## A. General information

1. Crosses in which parental individuals differ only in one trait and crosses in which only one trait is being considered are called ***monohybrid crosses***
2. For example, Mendel performed monohybrid crosses between tall pea plants and dwarf pea plants
   a. The resulting $F_1$ pea plants were all tall
   b. When Mendel crossed two $F_1$ pea plants together, the resulting $F_2$ generation contained both tall plants and dwarf plants
   c. The ratio of tall plants to dwarf plants in the $F_2$ population was 3 to 1

## B. Symbols and terminology

1. Although Mendel's $F_1$ pea plants contained factors from both tall and dwarf parents, all $F_1$ plants resembled the tall parent
   a. The tall phenotype therefore is ***dominant*** to the dwarf phenotype
   b. The dwarf phenotype therefore is ***recessive*** to the tall phenotype
   c. The parental phenotypes did not blend together to yield medium-sized $F_1$ plants
2. Because the dwarf phenotype reappears in the $F_2$ generation, the factor responsible for dwarf plants must be maintained — in an unaltered form — in the $F_1$ generation
   a. Each $F_1$ individual must inherit a tall factor from one parent and a dwarf factor from the other parent
   b. Each $F_1$ individual must therefore contain two factors
   c. If the number of factors is to remain constant from one generation to the next, *gametes* (male and female reproductive cells) must contain only one factor
3. The entity that controls heritable traits now is called a ***gene***
   a. Every individual possesses two genes for each trait
   b. Gametes contain only one of the two possible genes
   c. Genes may exist in alternate forms
      (1) The height gene of the pea, for example, may dictate a tall plant or a dwarf plant
      (2) Specific forms of a gene are called ***alleles***
      (3) The term allele is synonymous with Mendel's term "factor"
      (4) Alleles that confer dominant phenotypes are dominant alleles; alleles that confer recessive phenotypes are recessive alleles
4. Rules of genetic nomenclature dictate that genes are designated by italicized lowercase and uppercase letters
   a. Gene symbols typically are assigned on the basis of a ***mutant*** (deviant) phenotype
   b. Recessive alleles are designated by italicized lowercase letters; for example, *d* represents the dwarf allele (in this case, the dwarf phenotype is mutant)
   c. Dominant alleles are designated by italicized uppercase letters; for example, *D* represents the tall allele (in this case, the tall phenotype is normal)
      (1) The normal form of a trait is called the ***wild-type*** phenotype
      (2) By extension, the genotype that confers the wild-type phenotype is the wild-type genotype

d. The same letter is used for both alleles to indicate that they are alternate forms of the same gene (rather than *d* for dwarf and *T* for tall)
e. Genes also may be symbolized by two, three, or additional letters
f. To simplify gene nomenclature, a wild-type allele may have a + following its italicized letters (for example, *D+* )

5. ***Diploid*** organisms, which contain two sets of genes, may contain two dominant alleles (*DD*), two recessive alleles (*dd* ), or one dominant allele and one recessive allele (*Dd* )
   a. In ***homozygous*** organisms, or homozygotes, both alleles are the same
      (1) *DD* individuals are homozygous dominant (*DD* homozygotes)
      (2) *dd* individuals are homozygous recessive (*dd* homozygotes)
   b. In ***heterozygous*** organisms, or heterozygotes, the two alleles are different
      (1) Heterozygotes express the dominant allele
      (2) They are called ***carriers*** because they carry, but do not express, the recessive allele
   c. The symbol *D–* may be used to represent the genotype of a phenotypically dominant individual (*DD* or Dd )

## C. The principle of segregation

1. Homozygous individuals produce only one type of gamete
   a. DD individuals produce *D*-containing gametes
   b. *dd* individuals produce *d*-containing gametes
2. Heterozygous individuals produce two types of gametes
   a. *Dd* individuals produce both *D*-containing and *d*-containing gametes
   b. Gamete types are produced in equal number (50% of gametes contain *D* and 50% of gametes contain *d* )
3. Although different alleles may be combined in heterozygous individuals, the alleles can "disassociate" from one another during meiosis in the following generation
   a. This process is called ***segregation***
   b. It constitutes Mendel's first law
4. The results of segregation can be determined using a ***Punnett square***
   a. A Punnett square is a diagrammatic representation of the possible allele combinations resulting from the cross of two individuals
   b. Gametes produced by one parent are listed across the top of the square and gametes produced by the other parent are listed along the side of the square; the allele combinations that occur in the offspring are listed in the cells of the square
   c. Reexamination of Mendel's pea plant experiment shows that he crossed a pure-breeding tall plant (*DD*) with a pure-breeding dwarf plant (*dd* ) and obtained tall $F_1$ plants (*Dd* )
   d. The $F_1$ plants produce two types of gametes: *D* and *d*
   e. Two *D* gametes may unite to form a *DD* individual, two *d* gametes may combine to form a *dd* individual, or a *D* gamete may combine with a *d* gamete to form a *Dd* individual
   f. The individual cells of the Punnett square represent these possible gene combinations
5. Segregation results in three $F_2$ genotypes: homozygous dominant (*DD*), heterozygous (Dd and *dD* ), and homozygous recessive (*dd* )
   a. Of the $F_2$ individuals, 25% are *DD,* 50% are *Dd,* and 25% are *dd*
   b. The $F_2$ genotypic ratio is 1:2:1

**Monohybrid Cross**

Segregation of the *G* and *g* alleles results in the formation of different $F_2$ genotypes.

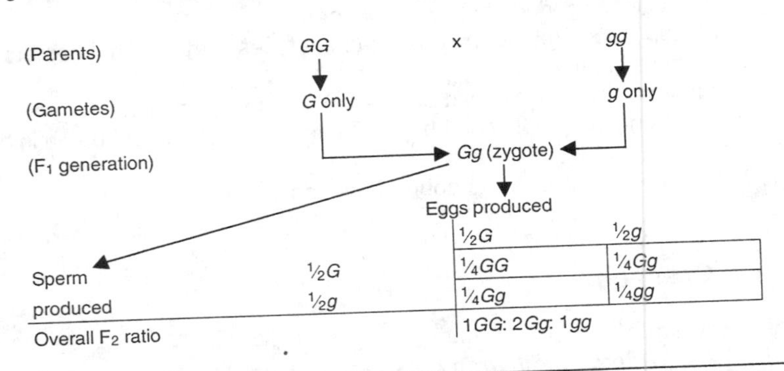


From *An Introduction to Genetic Analysis* (5th ed.) by Suzuki, Griffiths, Miller, Lewontin, and Gelbart. ©1993 by W.H. Freeman and Company. Reprinted with permission.

6. Segregation results in two $F_2$ phenotypes: tall (*DD* and *Dd*) and dwarf (*dd*)
   a. Seventy-five percent of $F_2$ individuals are tall and 25% are dwarf
   b. The $F_2$ phenotypic ratio is 3:1
7. The genotypic and phenotypic ratios shown above are consistent with the ratios observed by Mendel (see *Monohybrid Cross*)

**D. The testcross**

1. Genotypes must be determined experimentally because heterozygous individuals are phenotypically identical to individuals that are homozygous dominant
2. A convenient way to determine the genotype of an individual that displays a dominant phenotype is to cross it with an individual known to be homozygous recessive
3. The cross of an individual of an unknown genotype with a homozygous recessive individual is called a ***testcross***
   a. If the unknown genotype is *Dd,* crossing with a *dd* individual will result in tall (*Dd*) and dwarf (*dd*) progeny (1:1 ratio)
   b. If the unknown genotype is *DD, crossing with a dd individual will result only in tall (Dd*) progeny
4. In testcross populations (unlike $F_2$ populations), the phenotype reveals the exact genotype
   a. Testcross progeny that exhibit the dominant phenotype are heterozygous
   b. Testcross progeny that exhibit the recessive phenotype are homozygous recessive

**E. Problem-solving strategies**

1. The first step involves writing down what already is known
   a. Diagram all crosses, using standard genetic nomenclature and Punnett squares
   b. Make valid assumptions about genotypes

2. The second step involves the determination of the gamete types produced by the individuals being crossed
   a. Homozygotes produce only one type of gamete
   b. Heterozygotes produce two types of gametes
3. The third step is to combine maternal gamete types with paternal gamete types in all possible combinations
   a. A Punnett square format can be used in this determination
   b. The resulting genotypic and phenotypic ratios should be observed in the cross
4. The final step involves drawing appropriate conclusions

## III. Dihybrid Crosses

### A. General information

1. Mendel also performed ***dihybrid crosses,*** in which the parental generation differed by two genes, with each gene controlling a separate trait
   a. *RRyy* plants had seeds that were round (dominant allele *R*) and green (recessive allele *y*)
   b. *rrYY* plants had seeds that were wrinkled (recessive allele *r*) and yellow (dominant allele *Y*)
2. $F_1$ seeds of a dihybrid cross expressed the dominant alleles of each gene pair; thus, they were round and yellow (*RrYy*)
3. Four types of $F_2$ seeds were observed: round and yellow, round and green, wrinkled and yellow, and wrinkled and green
4. The ratio of $F_2$ phenotypes was 9:3:3:1

### B. The principle of independent assortment

1. When examined separately, individual traits of a dihybrid cross still exhibit the ratios observed in the $F_2$ progeny of a monohybrid cross
   a. Mendel's ratio of round to wrinkled seeds was 3:1
   b. Mendel's ratio of yellow to green seeds was 3:1
2. A ratio of 9:3:3:1 in the $F_2$ progeny can be viewed as a combination of two independent 3:1 ratios
   a. The gametes produced by the dihybrid $F_1$ plant are *RY, Ry, rY,* and *ry*
   b. Each type of gamete is equally likely to be produced because alleles of one gene pair segregate independently of alleles of other gene pairs during meiosis
      (1) This process is called ***independent assortment***
      (2) It constitutes Mendel's second law
   c. A Punnett square diagram for a dihybrid cross of $F_1$ plants consists of four rows and four columns (see *Dihybrid Cross*)
      (1) Nine cells of the square symbolize $F_2$ individuals that express the dominant allele of both genes (round and yellow)
      (2) Three cells of the square symbolize individuals that express the dominant allele of the first gene pair and the recessive allele of the second gene pair (round and green)
      (3) Three cells of the square symbolize individuals that express the recessive allele of the first gene pair and the dominant allele of the second gene pair (wrinkled and yellow)

## Dihybrid Cross

Segregation of the round (*R*) and wrinkled (*r*) alleles, along with the green (*y*) and yellow (*Y*) alleles, results in the formation of different $F_2$ genotypes and phenotypes.

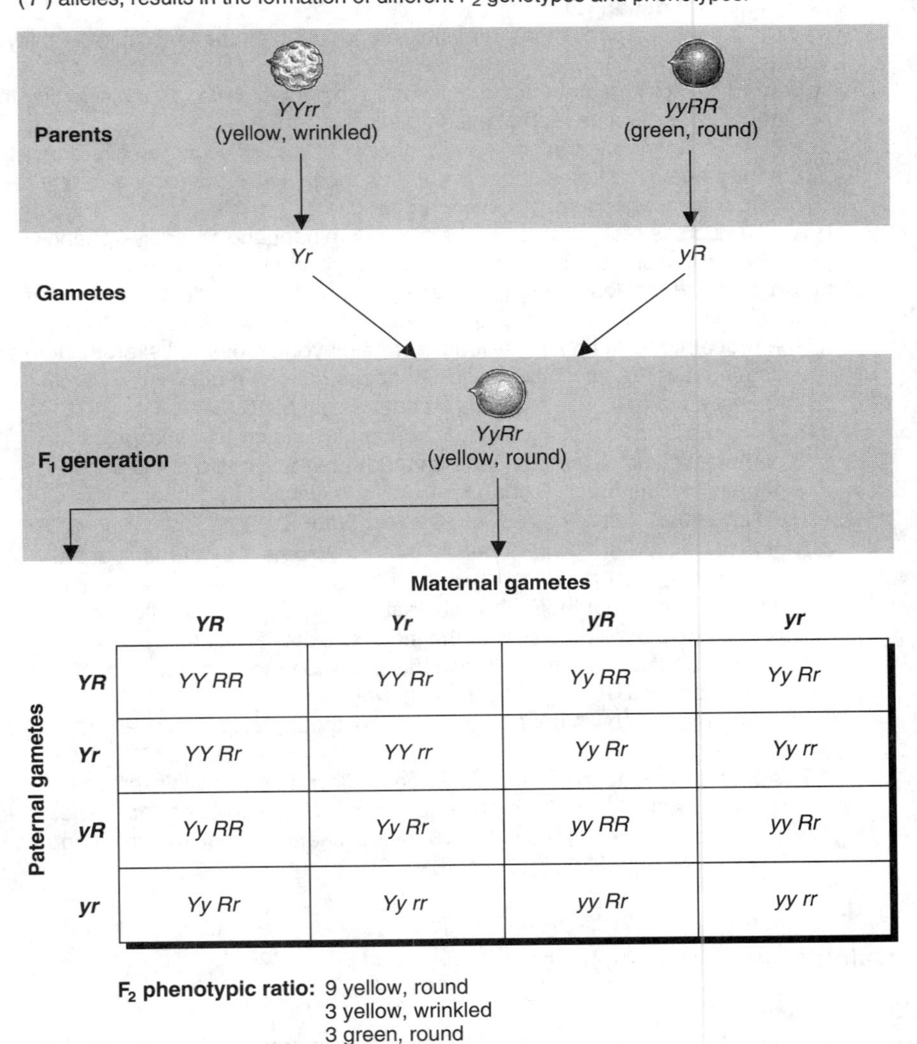


| Paternal \ Maternal | **YR** | **Yr** | **yR** | **yr** |
|---|---|---|---|---|
| **YR** | *YY RR* | *YY Rr* | *Yy RR* | *Yy Rr* |
| **Yr** | *YY Rr* | *YY rr* | *Yy Rr* | *Yy rr* |
| **yR** | *Yy RR* | *Yy Rr* | *yy RR* | *yy Rr* |
| **yr** | *Yy Rr* | *Yy rr* | *yy Rr* | *yy rr* |

**$F_2$ phenotypic ratio:** 9 yellow, round
3 yellow, wrinkled
3 green, round
1 green, wrinkled

(4) One cell of the square symbolizes the individual that expresses the recessive alleles of both gene pairs (wrinkled and green)

3. A dihybrid testcross ratio is a combination of the two independent monohybrid ratios
   a. If a heterozygous individual (*RrYy*) is crossed to a homozygous recessive (*rryy*), the following four genotypes — each in equal proportion — would result: *RrYy, rrYy, Rryy,* and *rryy*

b. A dihybrid testcross ratio of 1:1:1:1 is a combination of the two independent 1:1 ratios

### C. Probability and Mendelian ratios

1. The calculation of genetic ratios is easier to understand if the laws of probability are applied
2. ***Probability*** is the number of times an event is expected to happen, divided by the number of opportunities for that event to occur
    a. A 3:1 $F_2$ ratio means that the dominant phenotype will occur, on average, at a frequency of 3 times out of every 4 chances, while the recessive phenotype will occur once out of every 4 chances
    b. A 1:1 testcross ratio means that both the dominant and recessive phenotypes will occur, on average, 1 out of every 2 chances
3. The ***product rule*** states that the probability that two independent events will occur simultaneously is equal to the product of the individual probabilities
    a. The probability that an $F_2$ individual from a dihybrid cross will express both dominant traits is calculated as: 3 chances out of 4 multiplied by 3 chances out of 4, which equals 9 chances out of 16 ($3/4 \times 3/4 = 9/16$)
    b. The probability that an $F_2$ individual from a dihybrid cross will express the dominant allele of the first gene and the recessive allele of the second gene is calculated as: 3 chances out of 4 multiplied by 1 chance out of 4, which equals 3 chances out of 16 ($3/4 \times 1/4 = 3/16$)
    c. A ratio of 9:3:3:1 for $F_2$ individuals can be viewed as the net result of four probabilities: 9/16, 3/16, 3/16, and 1/16
4. The ***sum rule*** states that the probability that one of several mutually exclusive events will take place is the sum of the individual probabilities
    a. Mutually exclusive events are those that preclude the occurrence of other events; this also is known as the either/or rule
    b. For example, a baby is either a boy or a girl; these designations are mutually exclusive
    c. The probability that an $F_2$ individual from a dihybrid cross will express either the dominant allele of both genes or the recessive allele of both genes is calculated as: 9 chances out of 16, plus 1 chance out of 16, which equals 10 chances out of 16 ($9/16 + 1/16 = 10/16$)

## IV. Calculating Genetic Ratios

### A. General information

1. Although Punnett squares are useful in diagramming monohybrid and dihybrid crosses, they become less useful as the number of segregating genes increases
2. A second diagrammatic tool is the branch diagram, or forked-line diagram
3. A third method for calculating genetic ratios relies entirely on mathematical manipulation of probabilities

### B. Branch diagrams

1. A branch diagram graphically illustrates alternative phenotypes (or genotypes) for every segregating allele pair, thereby illustrating all possible combinations (see *Branch Diagram*)

### Branch Diagram

A branch diagram, shown below, can be used to calculate $F_2$ phenotypic ratios in a dihybrid cross. Dihybrid phenotypic proportions are calculated as the product of individual monohybrid proportions.

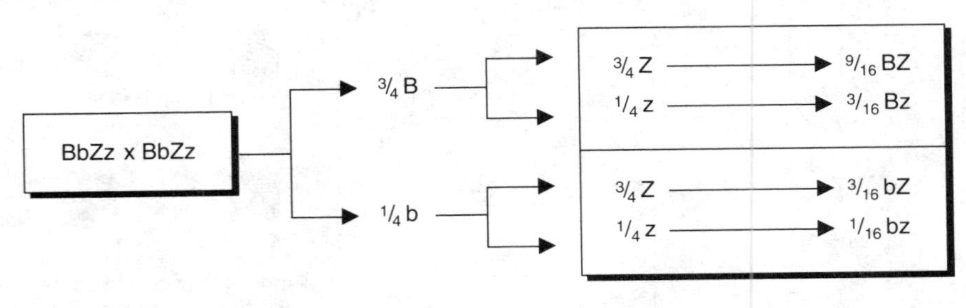


2. Genetic ratios are calculated by multiplying together individual probabilities that occur along the same line (applying the product rule)

**C. Probability**

1. Diagrammatic methods for ratio determination become impractical when segregating the alleles of many genes
2. This becomes obvious, for example, when one must determine what proportion of progeny from the cross of *AABbCCDdee* and *AaBbCcDdee* will be *AABBCcddee*
   a. One-half of the progeny will be *AA,* one-fourth of the progeny will be *BB,* one-half of the progeny will be *Cc,* one-fourth of the progeny will be *dd,* and all of the progeny will be *ee*
   b. The proportion of progeny that will be *AA* and *BB* and *Cc* and *dd* and *ee* is calculated as: $\frac{1}{2} \times \frac{1}{4} \times \frac{1}{2} \times \frac{1}{4} \times 1 = 1/64$

## V. Human pedigree analysis

**A. General information**

1. Because controlled matings do not occur in human populations, geneticists rely on records of previous matings (along with the associated phenotypes of the persons involved); this record is called a ***pedigree***
2. Pedigrees typically are initiated by geneticists, in conjunction with a *propositus* (a person seeking genetic advice)

**B. Performing a pedigree analysis**

1. Standard symbols have been adopted for use in human pedigree analysis (see *Pedigree Analysis, page* 14)
2. The first step in studying a completed diagram is to determine the mode of inheritance for the trait in question; that is, determining whether the trait is dominant or recessive
   a. Individuals who express a dominant trait must have at least one affected parent

## Pedigree Analysis

Standard symbols used in human pedigrees are shown below, along with an example of a pedigree that illustrates inheritance of a recessive disorder. Note that some terms will be discussed in later chapters, but are included here for completeness.

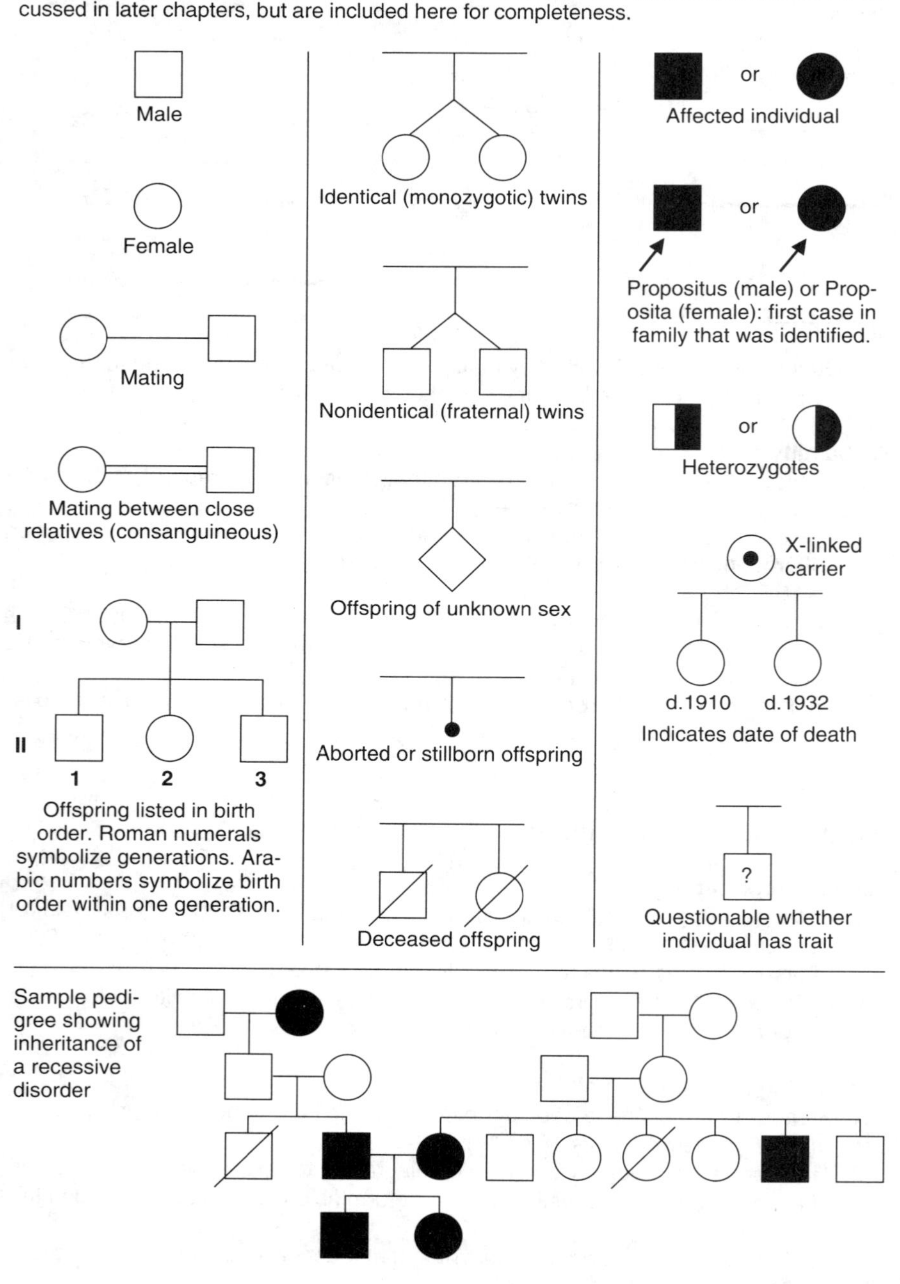

   b. Individuals who express a recessive trait must be homozygous and may or may not have affected parents
3. The second step is the determination of as many genotypes as possible
4. Investigators may then answer questions of probability, such as "What is the probability that the propositus is a carrier of a recessive defect?" or "What is the probability that the propositus will have a child who expresses a particular genetic defect?"

---

## Study Activities

1. Mendel drew two key conclusions (now called Mendel's laws) after observing the inheritance of plant characteristics. What are these laws and how did Mendel draw his conclusions?
2. Achondroplasia is a form of human dwarfism caused by a single, mutant gene. Two individuals with achondroplasia marry and have two children, one dwarf and one normal. Determine the mode of inheritance for achondroplasia and deduce the genotypes of each person described above.
3. You are given a cow that expresses a black and white dominant color pattern. However, some bulls and cows from the same family exhibit a red and white recessive color pattern. How might you determine if your cow is homozygous dominant or heterozygous?
4. How many different types of gametes will be produced by Judy, who is heterozygous for albinism (recessive trait) and heterozygous for familial hypercholesterolemia (dominant trait)?
5. Judy marries a man of the same genotype. What proportion of their children will be albino and exhibit hypercholesterolemia?
6. Draw two human pedigrees, one that illustrates transmission of a recessive trait and one that illustrates transmission of a dominant trait, across several generations.

# 3

# Advanced Mendelian Analysis

## Objectives

After studying this chapter, the reader should be able to:
- Distinguish between full dominance, incomplete dominance, and codominance.
- Explain the difference between segregation of multiple alleles and segregation of multiple genes.
- Describe the modifications of Mendelian ratios caused by epistasis, complimentary gene action, duplicate gene action, and lethal genes.
- Give an example of pleiotropy.
- Describe instances of variable penetrance and variable expressivity.

## I. Spectrum of Dominance

### A. General information

1. Although the alleles described in Chapter 2 were fully dominant or fully recessive, this condition is not dictated by, nor required for, Mendel's laws
2. Alleles may show several levels of dominance, including incomplete dominance

### B. Incomplete dominance

1. ***Incomplete dominance*** (also called semi-dominance) is the condition in which the phenotype of the heterozygote is intermediate to both homozygous genotypes
   a. In the four-o'clock plant, *pure-breeding* (referring to identical organisms that always produce progeny like themselves) varieties may have red or white flowers
   b. When a red-flowered four-o'clock plant is crossed with a white-flowered one, the resulting $F_1$ generation has pink flowers
   c. In the $F_2$ generation, red flowers, pink flowers, and white flowers occur in a 1:2:1 ratio
   d. The 1:2:1 phenotypic ratio is consistent with the $F_2$ genotypic ratio observed by Mendel
2. Because neither the red-flower allele nor the white-flower allele is fully dominant, the standard uppercase and lowercase allele symbols are not applicable
   a. Geneticists commonly use subscripts or superscripts to indicate allelic forms of genes that display incomplete dominance
   b. We use the genotypes $C_1C_1$, $C_1C_2$, and $C_2C_2$ to symbolize the red, pink, and white phenotypes, respectively

3. An example of incomplete dominance in humans is the carrying of the sickle cell anemia allele by heterozygotes
   a. Homozygous normal individuals (*$Hb^AHb^A$*) have normal red blood
   b. Homozygous individuals with sickle cells (*$Hb^SHb^S$*) have abnormal, sickle-shaped red blood cells
   c. Heterozygous individuals (*$Hb^AHb^S$*) have red blood cells that may exhibit the characteristic sickle shape under some circumstances

### C. Codominance

1. ***Codominance*** is the condition in which the heterozygote displays both homozygous phenotypes simultaneously
   a. Humans who are homozygous for the allele $I^A$ have type A blood and possess antigen A
   b. Humans who are homozygous for the allele $I^B$ have type B blood and possess antigen B
   c. Humans who are heterozygous for the alleles $I^A$ and $I^B$ have type AB blood and possess antigens A and B
2. $F_2$ populations resulting from a cross that illustrates codominance consist of individuals who express the phenotype of the first allele, both alleles, or the second allele in a 1:2:1 ratio

### D. Dominance relationships

1. Dominance relationships for any allele pair can be examined on the molecular, cellular, or organismal level
   a. We have learned that individuals with the genotype $Hb^AHb^S$ have red blood cells that exhibit incomplete dominance for cell shape
   b. However, at the molecular level, $Hb^AHb^S$ individuals contain two forms of hemoglobin
      (1) Both alleles give rise to hemoglobin molecules that differ according to function
      (2) Thus, at the molecular level, the alleles demonstrate codominance
   c. At the organismal level, $Hb^AHb^S$ individuals do not have anemia, as do $Hb^SHb^S$ individuals; therefore, the $Hb^A$ allele is fully dominant at this level
2. Dominance relationships apply specifically to the alleles being considered and may not apply to all of the alleles at a given ***locus***
   a. Some genes have more than two forms at a given locus and thus give rise to *multiple alleles*
   b. Human blood type is determined not only by the alleles $I^A$ and $I^B$, but also by the allele *i*
   c. Heterozygous individuals of genotype $I^Ai$ do not exhibit codominance, as do $I^AI^B$ individuals; instead, $I^A$ is fully dominant in genotype $I^Ai$

## II. Multiple Alleles

### A. General information

1. Some genes have more than two possible alleles; the set of alleles for any one gene is called the *allelic series*
2. In a diploid organism, however, only two alleles can be present in an individual at one time

### B. The allelic series

1. The ABO blood grouping used to illustrate codominance also serves as an example of multiple alleles
   a. An individual's genotype may be $I^AI^A$, $I^BI^B$, *ii*, $I^AI^B$, $I^Ai$, or $I^Bi$
   b. $I^A$ and $I^B$ are fully dominant when linked to *i*, yielding type A and type B blood, respectively
   c. $I^AI^B$ heterozygotes have type AB blood
   d. Homozygotes with the genotype *ii* have type O blood
2. A second example of multiple alleles are the gene forms that control the color of rabbit fur
   a. The alleles $c^+$, $c^{ch}$, $c^h$, and *c* are responsible for full-color (wild-type), chinchilla, Himalayan, and albino rabbits, respectively
   b. The order of alleles is, from most dominant to least dominant, $c^+$, $c^{ch}$, $c^h$, and *c*
   c. A cross of a wild-type rabbit of genotype $c^+c^h$ with a chinchilla rabbit of genotype $c^{ch}c$ will yield chinchilla rabbits of genotype $c^{ch}c^h$, wild-type rabbits of genotype $c^+c^{ch}$, wild-type rabbits of genotype $c^+c$, and Himalayan rabbits of genotype $c^hc$
   d. The phenotypic ratio of this cross is 1 chinchilla: 2 wild-type: 1 Himalayan

### C. The allelism test

1. The allelism test can help determine if different phenotypes are caused by multiple alleles of a single gene or by multiple genes
2. If individuals of different pure-breeding lines are crossed in all possible combinations and the ensuing $F_2$ progeny display monohybrid phenotypic ratios (3 dominant: 1 recessive), then the contrasting phenotypes are determined by alleles of the same gene
3. If individuals of different pure-breeding lines are crossed in all possible combinations and the ensuing $F_2$ progeny do not display phenotypic monohybrid ratios, then the contrasting phenotypes are determined by more than one gene

## III. Gene Interaction

### A. General information

1. Many traits are not determined by alleles of a single gene; in reality, the ultimate phenotype depends upon an interaction of two or more genes
2. Coat color in mice is determined by at least five genes: *A*, *B*, *C*, *D*, and *S*
   a. The *A* gene determines the presence or absence of a band of yellow color on individual hairs
      (1) Expression of the dominant *A* allele causes the yellow band, resulting in the wild-type pattern (called agouti)
      (2) Expression of the recessive *a* allele results in completely black hairs, thus giving a mouse an overall black phenotype
   b. The *B* gene determines the background color of the hair
      (1) Expression of the dominant *B* allele results in a black background
      (2) Expression of the recessive *b* allele results in a brown background
         (a) *A-B-* mice are agouti (yellow band; black hair)
         (b) *aaB-* mice are black (no yellow band; black hair)
         (c) *A-bb* mice are cinnamon (yellow band; brown hair)

(d) *aabb* mice are brown (no yellow band; brown hair)

c. The *C* gene controls total pigment production

(1) Expression of the dominant *C* allele results in full color expression of genes *A* and *B*

(2) Expression of the recessive allele *c* results in absence of pigment, producing an albino mouse regardless of the allelic composition of the *A* and *B* genes

d. The *D* gene determines the intensity of color

(1) Expression of the dominant *D* allele results in full-color intensity

(2) Expression of the recessive allele *d* results in weak, or dilute, color

e. The *S* gene controls the absence or presence of spotting

(1) Expression of the dominant *S* allele results in no spotting

(2) Expression of the recessive allele *s* results in white spotting that is superimposed on any of the previously mentioned colors

3. The $F_2$ progeny resulting from the crossing of phenotypically distinct, pure-breeding lines of mice will not always yield monohybrid ratios; this indicates that two or more gene pairs are segregated

**B. Epistasis**

1. A consideration of the coat color genes of mice leads to the conclusion that the *C* gene controls the ultimate deposition of pigment in mice
2. The situation in which one gene prevails over a second, nonallelic, gene is called ***epistasis***
3. The *C* gene therefore is epistatic to the other pigment-controlling genes; according to the literal meaning of the word epistasis, the *cc* genotype "stands upon" all other gene combinations
4. Segregation of an epistatic gene, along with other genes, gives rise to modified Mendelian ratios

**C. Complementary gene action**

1. In some two-gene interactions, specific phenotypic expression requires that dominant alleles of both gene pairs be present
2. Because the genes are said to complement one another, this type of gene action is called ***complementary gene action***
3. Recessive homozygosity in either gene will prevent the specific phenotype from occurring
   a. In corn, *A1* and *A2* are two of several genes required for kernel pigmentation
   b. *A1-A2-* kernels are fully pigmented, whereas *A1-a2a2, a1a1A2-,* and *a1a1a2a2* kernels lack pigment
4. As with epistatic gene action, complementary gene action results in modified Mendelian ratios (a 9:7 ratio, for example)

**D. Duplicate gene action**

1. In some two-gene interactions, specific phenotypic expression requires that the dominant allele of only one gene pair be present
2. Because the genes are said to duplicate one another, this type of action is called *duplicate gene action*
3. A dominant allele at the first gene, the second gene, or at both genes will result in the specific phenotype

4. Duplicate gene action results in modified Mendelian ratios (a 15:1 ratio, for example)

## IV. Lethal Alleles

### A. General information

1. The genetic variants discussed above do not have a marked effect on the viability of organisms
2. In the many genes that are essential for growth and development, however, genetic variation may result in ***lethal alleles***
3. Lethal alleles may exert their effects at various phases of development, including the embryonic stage, childhood, and adulthood
4. Segregation of lethal alleles in a population may produce nonviable progeny
   a. The tailless condition of Manx cats is conditioned by the $M^L$ allele
      (1) Cats homozygous for the wild-type *M* allele develop tails
      (2) Heterozygous cats ($M^LM$) are tailless as a result of slightly abnormal spinal development
      (3) Cats homozygous for $M^L$ do not develop past the embryonic stage, due to severe spinal abnormalities
   b. Examples of lethal alleles in humans include those responsible for Duchenne muscular dystrophy, cystic fibrosis, and Huntington's disease

### B. Genetic behavior

1. Lethal alleles that prevent development of an organism past the embryonic stage may lead to unusual segregation ratios among viable progeny
   a. Viable progeny resulting from the cross of two Manx cats will be tailless ($M^LM$) or normal ($MM$), and will occur in a 2:1 ratio
   b. Viable progeny resulting from the cross of a Manx cat and a normal cat will be tailless ($M^LM$) or normal ($MM$), and will occur in a 1:1 ratio
2. Phenotypes resulting from heterozygosity of a lethal allele will not be pure-breeding

## V. Pleiotropy

### A. General information

1. Variation in a single gene often gives rise to variability in many seemingly unrelated traits
2. The association of several unrelated phenotypes with a single ***mutation*** is called ***pleiotropy***
3. Genes that influence more than one trait are called *pleiotropic genes*

### B. Examples in humans

1. Homozygosity for the sickle cell anemia allele results in anemia, physical weakness, hypertrophy of bone marrow, skin ulcers, restricted blood flow, and susceptibility to heart failure and rheumatism
2. Marfan syndrome, which is caused by a dominant allele, is associated with a weakened cardiovascular system, tall and thin body type, and vision defects

3. Homozygosity for the cystic fibrosis allele results in reduced digestive efficiency, chronic lung infections, and abnormal functioning of exocrine glands in the skin, lungs, pancreas, and liver
4. Pleiotropic genes give rise to an array of phenotypic abnormalities
5. A condition characterized by multiple phenotypic abnormalities (of genetic or nongenetic origin) is called a syndrome

## VI. Penetrance and Expressivity

### A. General information

1. Phenotype expression is variable for some genes
2. Environmental and developmental effects may alter gene expression
3. The presence of a gene does not always result in an observable phenotypic effect
   a. The proportion of individuals with a specific genotype that express the corresponding phenotype is called ***penetrance***
   b. Such a gene is said to show incomplete penetrance
4. The presence of a gene does not always result in the same degree of phenotypic effect
   a. The degree of phenotype expression of a particular genotype is termed ***expressivity***
   b. Such a gene is said to show variable expressivity

### B. Examples in humans

1. Camptodactyly, which is caused by a dominant allele, is characterized by permanently flexed fingers or toes
   a. Because not every individual carrying this allele will show the distinctive phenotype, the gene for camptodactyly exhibits incomplete penetrance
   b. Because individuals who express the phenotype may have one or more fingers or toes affected, the gene exhibits variable expressivity
2. The dominant allele responsible for retinoblastoma, an eye tumor, also shows incomplete penetrance because not all carriers of the allele develop the tumor

## Study Activities

1. A male of blood type M (containing antigen M) and a female of blood type MN (containing antigens M and N) have five offspring, three with blood type M and two with blood type MN. Assuming that a single gene is responsible for this trait, determine the dominance relationships for the alleles of this gene and assign appropriate genotypes to each person.
2. Palomino horses are not the result of pure breeding. When mated to each other, palominos will yield 25% cremellos (nearly white), 50% palominos, and 25% chestnuts. Determine the mode of inheritance for this trait.
3. Two persons with albinism marry and have children who are all normal. What are the probable genotypes for the parents and the children?
4. Draw a Punnett square to illustrate the progeny produced when a chinchilla rabbit of genotype $c^{ch}c^{h}$ is crossed with an albino rabbit of genotype *cc*.

5. A corn mutant that contained elevated amino acid levels in the grain was discovered. Breeders, however, were unable to develop a pure-breeding line with elevated amino acids. Each time two high-amino acid plants were crossed, normal plants appeared 33% of the time. What would account for these results?
6. Several false statements are included in the following description of a genetic disease. Rewrite the following paragraph so that the information is correct. "Osteogenesis imperfecta is a disease characterized by fragile bones, deafness, and abnormal eye color. Because the gene affects multiple traits, it is called an incomplete penetrance gene. Individuals may show any one of the symptoms or a combinations of symptoms; therefore, the gene exhibits variable penetrance. Occasionally, individuals do not develop any abnormalities; therefore, the gene also exhibits variable expressivity."

# 4

# The Chromosome Theory of Inheritance

## Objectives

After studying this chapter, the reader should be able to:

- Sketch the cell cycle.
- Describe each stage of mitosis and meiosis.
- Compare and contrast mitosis and meiosis.
- List the similarities that exist between the behavior of chromosomes and the segregation of genes during meiosis.
- Explain how geneticists confirmed the chromosome theory of inheritance.
- Describe the chromosomal basis of sex determination.
- Identify characteristic features of sex-linked inheritance.
- Compare and contrast the diploid life cycle, the haploid life cycle, and the alternation of generations.

## I. The Eukaryotic Cell Cycle

### A. General information

1. Unlike prokaryotes, eukaryotes contain cells in which the genetic material is sequestered in a membrane-bound organelle called the nucleus
   a. The genetic material of eukaryotes is more complex than that of prokaryotes
   b. The genetic material ***deoxyribonucleic acid (DNA)*** is contained within ***chromosomes,*** the subcellular structures located in the eukaryotic cell nucleus
   c. In 1888, W. Waldeyer coined the term chromosome, which literally means "colored body," to describe the dark-staining nuclear structures that can be seen with a light microscope
2. *Diploid* organisms contain two sets of ***homologous chromosomes;*** this means that the chromosomes have the same structure and function
   a. The diploid number of chromosomes is represented by 2*n*
   b. In humans, the 2*n* chromosome number is 46, comprising two sets of 23 chromosomes
   c. Although chromosomes were discovered in the 1880s, the exact number of human chromosomes was not determined until 1956
3. Cell division that is associated with organismal growth and differentiation must include the faithful replication and assortment of chromosomes as well as other subcellular structures
4. The period of cell nondivision is called ***interphase***

a. During interphase, the cell actively is engaged in metabolic functions that are specific to that type of cell
b. Interphase is divided into three subphases: gap 1 ***($G_1$ phase),*** synthesis ***(S phase),*** and gap 2 ***($G_2$ phase)***
(1) During the $G_1$ phase, the cell prepares for DNA synthesis and chromosome replication
(2) The cell synthesizes DNA and replicates chromosomes during the S phase
(a) The products of chromosome replication are called *sister chromatids*
(b) A ***chromatid*** is one-half of a replicated chromosome
(c) Sister chromatids remain attached at the ***centromere,*** the constricted region of the chromosome that is involved in chromosome movement
(3) During the $G_2$ phase, the cell prepares for nuclear division and, typically, cytoplasmic division

5. The division phase of cells that are destined not to become gametes is called ***mitosis*** (M phase)
a. Nuclear division is called *karyokinesis*
b. Cell division is called *cytokinesis*
c. Mitosis generally is the shortest phase of the eukaryotic cell cycle; most of a cell's life is spent in the $G_1$ phase (see *Eukaryotic Cell Cycle*)
6. The division phase of cells that are destined to become gametes is called ***meiosis***

**B. Mitosis**

1. The process of mitosis ensures that daughter cells contain the same number of chromosomes as the parent cell
a. The number of chromosomes in somatic, or body, cells is designated as $2n$
b. Therefore, the parent cell and daughter cells each have a chromosome number of $2n$
2. Mitosis can be separated into four distinct phases: prophase, metaphase, anaphase, and telophase
a. During ***prophase,*** the first phase of nuclear division, chromosomes become shorter and thicker
(1) By late prophase, the chromatids that comprise the replicated chromosomes can be seen with a light microscope because the chromatids are compressed and tightly coiled
(2) The nuclear membrane disappears
(3) The nucleolus, which is the site of ***ribosomal ribonucleic acid (rRNA)*** synthesis, also disappears
(4) The mitotic spindle apparatus, composed of microtubules, appears
(a) The microtubules, which are composed of cylinders of tubulin protein and are parallel to each other, point to each of the two cell poles
(b) The spindle is shaped by microtubule organizing centers, of which the centriole is an example
b. During ***metaphase,*** chromosomes migrate to the equatorial plane and attach themselves to the microtubules
(1) The microtubules then become attached to the centromere

## Eukaryotic Cell Cycle

A cell's life cycle has two main parts: interphase and mitosis. Interphase, which is when division is not taking place, includes the gap 1 ($G_1$) phase, the synthesis (S) phase, and the gap 2 ($G_2$) phase. Also known as the M phase, mitosis refers to nuclear division; this includes prophase, metaphase, anaphase, and telophase. The diagram below illustrates the life cycle of a eukaryotic cell.

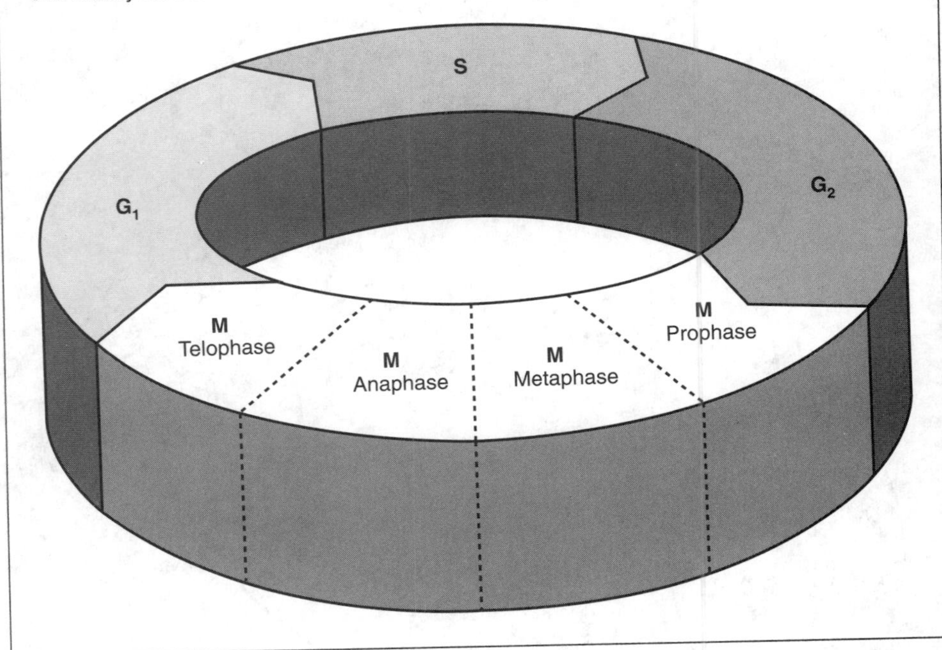


(2) The sister chromatids become attached to opposite cell poles

c. During ***anaphase,*** non–sister chromatids migrate to opposite poles

(1) Centromeric regions separate, facilitating the disjunction of sister chromatids

(2) Chromosomes are pulled to the poles by the microtubules attached to the centromeres

(3) In humans, 46 chromosomes migrate to each pole

d. ***Telophase*** begins when the chromosomes arrive at the cell poles

(1) Nuclear membranes reform around each daughter nucleus (which consists of one cell pole and the chromosomes attached to it)

(2) The nucleoli reappear

(3) Chromosomes then uncoil and elongate

3. Telophase typically is accompanied by cytokinesis

a. In animal cells, cytoplasmic division is accomplished by constriction (furrowing) of the cell, causing the two daughter nuclei to separate

b. In plant cells, a cell plate upon which new cell wall material is deposited forms between the two daughter nuclei

4. Diploid cells that undergo mitosis yield two identical, diploid daughter cells (see *Mitosis,* page 26)

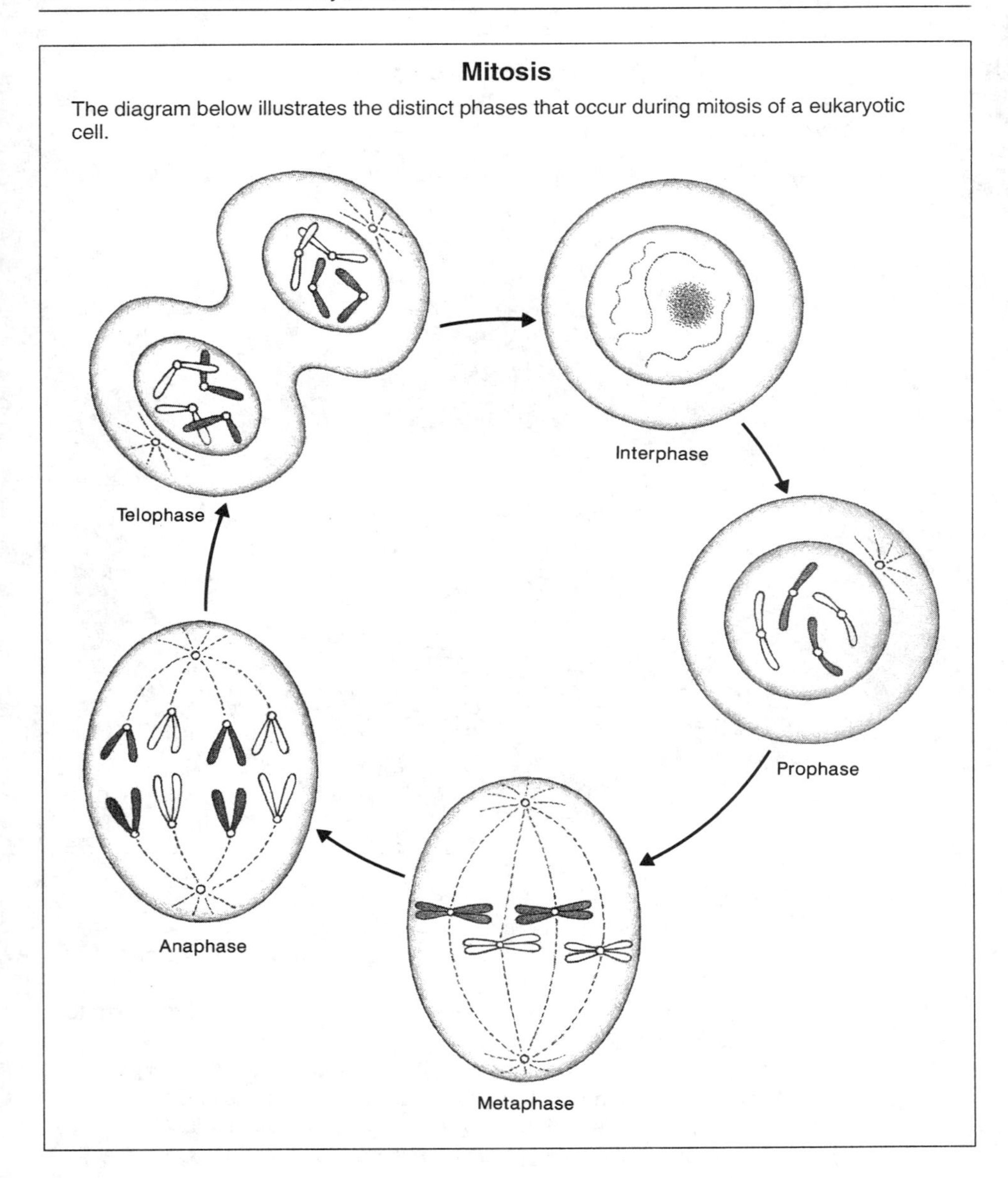


**Mitosis**

The diagram below illustrates the distinct phases that occur during mitosis of a eukaryotic cell.

### C. Meiosis

1. Meiosis is the process by which the number of chromosomes in cells that are destined to become gametes is reduced by one-half
   a. Cells that contain half the normal number of chromosomes are called ***haploid***
   b. The haploid number of chromosomes is represented by *n*
2. Meiosis is necessary in organisms that are the progeny of two parents; the nuclear fusion that occurs during fertilization must be counterbalanced so that number of chromosomes remain constant over generations

3. The process of meiosis consists of two successive nuclear divisions, meiosis I and meiosis II, which follow a single round of chromosome replication
   a. In diploid organisms, meiosis I reduces the number of chromosomes from diploid to haploid; this is called a *reduction division*
   b. Meiosis II results in the separation of sister chromatids; this is called an *equatorial division*
4. Meiosis can be separated into eight distinct phases: prophase I, metaphase I, anaphase I, telophase I, prophase II, metaphase II, anaphase II, and telophase II (see *Meiosis,* page 28)
   a. Prophase I, the most complex stage of meiosis, is further divided into five substages: leptonema, zygonema, pachynema, diplonema, and diakinesis
      (1) During leptonema, chromosomes first become visible as very long, thin threads (as seen under a light microscope)
      (2) During zygonema, the two sets of homologous chromosomes pair with one another in a process called ***synapsis***
         (a) Each chromosome has a pairing partner, or ***homolog***
         (b) Homologs are held together by a complex protein structure, the synaptonemal complex
         (c) Each pair of homologs is called a *bivalent*
      (3) During pachynema, homologs become thick and fully synapsed
         (a) Thick, dark, beadlike structures called chromomeres appear on the chromosomes
         (b) Each pair of homologs has a distinctive pattern of chromomeres
      (4) During diplonema, the replicated nature of each homolog appears
         (a) Each homologous pair consists of four chromatids, or a ***tetrad***
         (b) Cross-shaped structures called ***chiasmata,*** which are formed when non–sister chromatids break and rejoin, also appear
      (5) During diakinesis, chromosomes reach their greatest level of compactness
   b. During metaphase I, each pair of homologs aligns itself on the cell's equatorial plane
   c. Chromosome movement takes place during anaphase I, with homologous pairs separating and being pulled to opposite poles along with their intact centromeres (unlike mitosis)
      (1) The two chromatids of each homolog remain attached to the chromosome
      (2) In humans, 23 chromosomes, each containing two chromatids, migrate to each pole
   d. During telophase I, the chromosomes arrive at the cell poles
      (1) Each telophase nucleus is haploid *(n)* because each contains half of the normal number of chromosomes
      (2) In some organisms, chromosomes uncoil slightly and the nuclear membranes reform; in others, chromosomes remain fully condensed and immediately enter meiosis II
   e. Prophase II resembles the mitotic prophase because chromosomes are fully condensed and each is composed of two chromatids
   f. During metaphase II, chromosomes migrate to the cell's equatorial plane and become attached to the microtubules
      (1) The microtubules become attached to the centromere
      (2) The sister chromatids become attached to opposite cell poles

## Meiosis

The illustration below traces the path of two pairs of homologous chromosomes through the distinct phases of meiosis. Maternal chromosomes are shown in black; paternal chromosomes, in white.

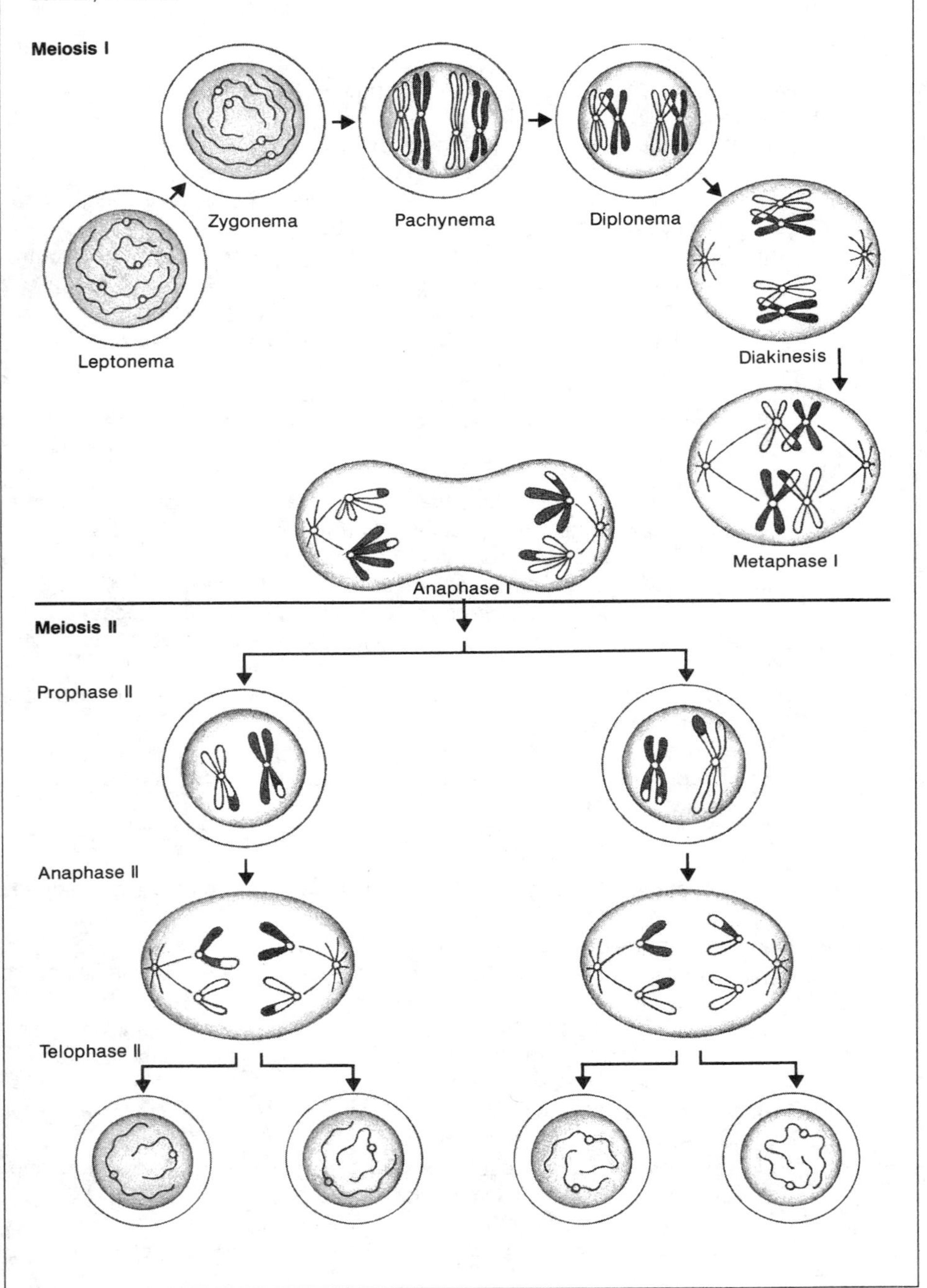

g. During anaphase II, sister chromatids separate and migrate to opposite poles
h. During telophase II, chromosomes arrive at the cell poles
(1) Nuclear membranes reform around each haploid daughter nucleus
(2) The nucleoli reappear
(3) Chromosomes uncoil and elongate
(4) Cells that undergo meiosis yield four haploid daughter cells

## II. Continuity Across Generations

### A. General information

1. The meiotic process is virtually identical in all organisms
2. However, the division of *meiocytes* (the cells undergoing meiosis) produces different products in plants and animals
   a. The direct products of animal meiotic division are haploid gametes (sperm and egg)
   b. The direct products of plant meiosis are haploid spores
3. Although Mendelian inheritance patterns pertain to all organisms that undergo meiosis, we must understand the differences between the life cycles of plants and animals before applying Mendel's laws

### B. Gametogenesis

1. The production of gametes in animals includes the formation of sperm (spermatogenesis) and eggs (oogenesis)
2. In spermatogenesis, a primary spermatocyte undergoes the first meiotic division to form two secondary spermatocytes; the two secondary spermatocytes then undergo a second meiotic division, giving rise to a total of four spermatids that mature into sperm
3. In oogenesis, a primary oocyte undergoes the first meiotic division to form one secondary oocyte and one polar body; the secondary oocyte undergoes a second meiotic division to form one ootid and one polar body, while the first polar body from meiosis I undergoes a second meiotic division to form two polar bodies
   a. Ootids mature to form eggs
   b. Unlike spermatogenesis, oogenesis produces only one gamete from each parent cell
   c. Asymmetrical cytokinesis results in one meiotic product that contains a large volume of cytoplasm (the egg) and three meiotic products containing very little cytoplasm (the polar bodies)

### C. Sporogenesis

1. In plants, diploid meiocytes undergo two meiotic divisions to form haploid spores
2. The haploid spores do not mature directly into gametes but divide mitotically, giving rise to a multicellular haploid stage of the life cycle
3. The haploid stage forms gametes via mitosis; these gametes fuse to re-form the diploid phase of the life cycle
4. Plants therefore exhibit an alternation of generations
   a. The diploid stage of the life cycle is called the ***sporophyte*** (spore-producing) stage

b. The haploid stage of the life cycle is called the ***gametophyte*** (gamete-producing) stage

**D. Other life cycles**

1. Meiosis in algal and fungal species also results in the formation of spores, which divide mitotically to form additional haploid cells
2. However, a variable pattern of generation alternation exists
3. In some organisms, the diploid phase of the life cycle is reduced so greatly that the only diploid cell is the zygote, which undergoes meiosis as soon as it is formed
4. Consequently, some organisms have a haploid life cycle rather than the diploid life cycle seen in animals and the alternation of generations that occurs in plants

## III. Parallel Behavior of Genes and Chromosomes

**A. General information**

1. In 1902, Walter Sutton and Theodor Boveri recognized that the segregation of Mendel's factors (alleles) was consistent with the segregation of chromosomes during meiosis
   a. Genes and chromosomes commonly occur in pairs
   b. Alleles and homologous chromosomes segregate equally into gametes
   c. Different genes and different homologous chromosome pairs segregate independently
2. Some people hypothesized that genes occur on chromosomes
3. The location of a gene on a chromosome is called its ***locus***
4. This idea served to unite the fields of cytology (cell biology) and genetics
5. Proof of the chromosome theory of inheritance was provided by Calvin Bridges in 1925, and was based on Thomas Morgan's studies of sex determination in *Drosophila* (fruit fly)

**B. Chromosomes and sex determination**

1. In 1905, Nettie Stevens noted that *Drosophila* contains four pairs of chromosomes
   a. One pair of chromosomes, which came to be designated X and Y, was noted to be heteromorphic, that is, containing homologs that were not identical
   b. Female *Drosophila* contained two X chromosomes, whereas male *Drosophila* contained one X and one Y
   c. Sex in *Drosophila* is determined by the ratio of X chromosomes (a ***sex chromosome***) to non-sex chromosomes (called ***autosomes***)
2. Morgan correlated the segregation of the X and Y chromosomes (both of these are sex chromosomes) of *Drosophila* with segregation of a mutant eye-color gene
   a. A white-eyed mutant male was crossed with a red-eyed (wild-type) female
      (1) Because the $F_1$ progeny had red eyes, the white-eyed trait was found to be recessive
      (2) Among the $F_2$ progeny, red-eyed and white-eyed flies occurred in a 3:1 ratio but all white-eyed flies were male

b. Morgan concluded, on the basis of the inheritance pattern, that the eye-color gene was located on the X chromosome and that the Y chromosome did not contain an eye-color gene
   (1) When white-eyed males (designated $X^{w}Y$) produce sperm, half the sperm contains an X chromosome ($X^{w}$) and the other half contains a Y chromosome (Y)
   (2) Red-eyed females ($X^{w+}X^{w+}$) produce eggs that contain an X chromosome ($X^{w+}$)
   (3) When these flies are crossed, the $F_1$ progeny are red-eyed males ($X^{w+}Y$) and red-eyed females ($X^{w}X^{w+}$)
   (4) $F_2$ generation flies are $X^{w+}X^{w}$ (red-eyed females), $X^{w+}X^{w+}$ (red-eyed females), $X^{w}Y$ (white-eyed males), and $X^{w+}Y$ (red-eyed males)

3. Note, however, that not all species that depend on chromosome prevalence for sex determination have females with an XX genotype and males with an XY genotype

**C. Proof of the chromosome theory**

1. The chromosome theory of inheritance was proven by the observation of abnormal segregation of the eye-color gene and the sex chromosomes in *Drosophila*
   a. Generally, crosses of white-eyed females ($X^{w}X^{w}$) and red-eyed males ($X^{w+}Y$) gave rise to red-eyed females ($X^{w+}X^{w}$) and white-eyed males ($X^{w}Y$)
   b. Sometimes red-eyed males and white-eyed females occurred as a result of this cross; such offspring are called exceptional progeny
   c. Geneticists discovered that these exceptional progeny resulted from ***nondisjunction*** (that is, nonseparation) of the sex chromosomes during meiosis
      (1) The white-eyed female parent sometimes produced eggs that carried two $X^{w}$ chromosomes or no $X^{w}$ chromosomes rather than one $X^{w}$ chromosome
      (2) Fertilization of these eggs by normal sperm produced $X^{w}X^{w}X^{w+}$ (nonviable), $X^{w}X^{w}Y$ (white-eyed female), $X^{w+}0$ (red-eyed male), and Y0 (nonviable)
2. The correlation between abnormal segregation of the eye-color gene and the abnormal segregation of the X chromosome proved that the chromosome theory of inheritance is correct (see *Segregation of Eye Color and Sex in Drosophila,* page 32)

**D. Chromosomal basis of Mendel's laws**

1. Mendel's laws are the direct reflection of chromosome behavior during meiosis
2. The movement of sex chromosomes and autosomes determines the genotypes of the resulting meiotic products, and, subsequently, the gametes
   a. Mendel's first law (segregation) is explained by the movement of homologous chromosomes to opposite poles during anaphase I of meiosis
   b. Mendel's second law (independent assortment) is explained by the random alignment of homologous chromosome pairs during metaphase I of meiosis

### Segregation of Eye Color and Sex in *Drosophila*

The diagram explains the segregation of eye color and sex in crosses of *Drosophila*.

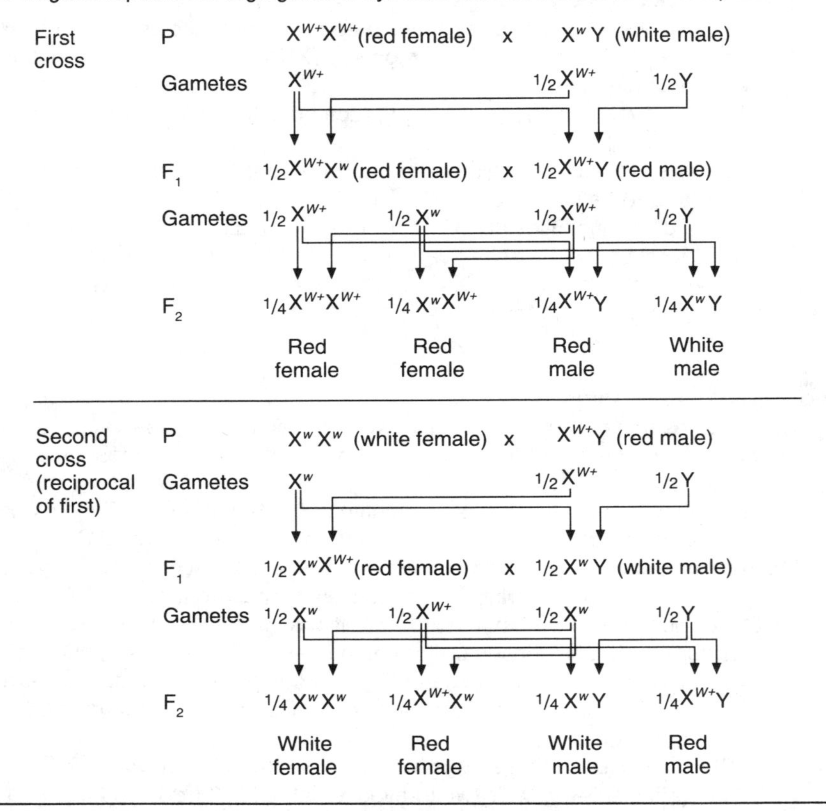


From *An Introduction to Genetic Analysis* (5th ed.) by Suzuki, Griffiths, Miller, Lewontin, and Gelbart.
©1993 by W.H. Freeman and Company. Reprinted with permission.

# IV. Sex Linkage

## A. General information

1. The behavior of genes located on sex chromosomes was described in and alluded to during our proof of the chromosome theory of inheritance in section II. above
2. Genes located on sex chromosomes exhibit ***sex linkage*** because segregation of these alleles is associated with the segregation of sex determination
3. Sex-linked traits may be further classified as X-linked recessive, X-linked dominant, and Y-linked

## B. X linkage

1. X-linked recessive traits require two recessive, X-linked alleles on a chromosome for expression in females (for example, the white-eye allele of *Drosophila*)

2. X-linked dominant traits require only one dominant, X-linked allele on a chromosome for expression in females
3. X-linked genes occur only in a single chromosome in males because the Y chromosome does not contain the same genes as the X
   a. Males cannot be homozygous or heterozygous for X-linked traits; instead, they are ***hemizygous*** (that is, "half-zygous" or having only one copy, or allele)
   b. Hemizygous alleles are expressed in males, whether they are dominant or recessive to other alleles
4. X-linked recessive inheritance has several distinct features
   a. More males than females express the recessive phenotype because females must inherit two of the same allele for expression but males need to inherit only one
   b. An affected father typically has daughters that carry the recessive allele; half the daughters' sons will exhibit X-linked recessive inheritance
   c. An affected father typically has sons that produce normal progeny (that is, they do not inherit the X-linked recessive trait)
5. X-linked dominant inheritance has several characteristics
   a. Affected males pass the X-linked trait to their daughters but not to their sons
   b. Heterozygous females pass the trait on to half their children, irrespective of gender
6. In humans, X-linked alleles are responsible for many disorders
   a. X-linked recessive traits include hemophilia, red-green color blindness, and Duchenne muscular dystrophy
   b. X-linked dominant traits are less common and include hyperphosphatemia, a metabolic disorder accompanied by deformities of the head, neck, trunk, and legs

**C. Y linkage**

1. Genes that occur on the Y chromosome (and are expressed only in males) are called Y-linked or holandric
2. Y-linked inheritance is characterized by consistent father-to-son transmission
3. The genes for male gender are the only human genes that have been proven to be Y-linked

---

## Study Activities

1. Draw a circular representation of the cell cycle. Label all phases, including those involving nuclear division. List the characteristic features of each stage.
2. Draw a diploid cell that contains six chromosomes: two long, two medium-length, and two short. Assuming that chromosomes of equal length are homologous, illustrate what the cell would look like in mitotic prophase, metaphase, anaphase, and telophase.
3. Make a second drawing of the cell described in question 2. Assuming that chromosomes of equal length are homologous, illustrate what the cell would look like in prophase I, metaphase I, anaphase I, telophase I, prophase II, metaphase II, anaphase II, and telophase II of meiosis.

4. Make another drawing of the cell described in question 2. Place the recessive allele *a* on one of the long chromosomes and the dominant *A* on the other. In a similar manner, make the medium-length chromosome pair heterozygous *Bb* and the short chromosome pair heterozygous for gene *Cc*. Draw the eight possible meiotic products.
5. If a red-green color-blind woman marries a man with normal vision, what are the possible genotypes and phenotypes of their children?
6. A man with a disorder of tooth enamel marries a phenotypically normal woman. They have eight children: five daughters inherited their father's faulty tooth enamel and three sons did not inherit the condition. How does this trait appear to be inherited?
7. A diploid plant of genotype *AaBb* undergoes meiosis and forms spores. The spores divide to form gametophytes, which yield gametes. If the gametes fertilize one another in random fashion, what will be the genotypic ratios in the ensuing sporophyte generation?

# 5

# Structure and Replication of Genetic Material

## Objectives

After studying this chapter, the reader should be able to:
- Describe how the chemical nature of genetic material was discovered.
- Discuss the chemical composition of nucleic acids.
- Describe the complementary and antiparallel nature of double-stranded deoxyribonucleic acid (DNA).
- Illustrate semiconservative replication by drawing progenitor and daughter DNA double helices.
- Draw a replication fork, showing regions of continuous and discontinuous DNA synthesis.
- List and describe the enzymatic functions that occur during replication.

## I. Genetic Material

### A. General information

1. Molecules that serve as genetic material possess specific characteristics
   a. They must contain all information required for cellular and organismal structure, function, development, and reproduction
   b. They must replicate accurately
   c. These molecules also must accommodate variability
2. The recognition of DNA as genetic material and knowledge of nucleic acid structure are relatively recent discoveries; Watson and Crick received a Nobel Prize in 1953 for their discovery of DNA structure
3. Bacterial ***transformation*** experiments, in which one genotype is transformed into another, were among the first studies to address the chemical nature of hereditary material
4. As a result of these experiments, genetic material was found to be DNA

### B. Bacterial transforming principle

1. In 1928, Frederick Griffith discovered that bacterial cells of one phenotype can be transformed into another phenotype
   a. Virulent strains of bacteria *(Streptococcus pneumoniae)* caused death when injected into mice
   b. Heat-killed virulent cells lost their ability to cause death
   c. Nonvirulent strains of bacteria did not cause death

d. When living, nonvirulent bacterial were grown in the presence of heat-killed virulent bacteria, the living cells were transformed into the virulent phenotype
e. Griffith concluded that a substance in the heat-killed cells caused transformation of the living cells

2. In 1944, Oswald Avery, Colin MacLeod, and Maclyn McCarty discovered that the transforming element is DNA
   a. Bacterial cell debris was separated into different types of molecules
   b. They found that only DNA could transform cells from one phenotype into another

### C. Genetic material of bacterial viruses

1. In 1952, Alfred Hershey and Martha Chase determined that the chemical responsible for heredity in *bacteriophage* (bacterial virus, or phage) T2 was DNA
   a. Bacteriophage T2 particles were alternately labeled with radioactive phosphorus and radioactive sulfur (sulfur is present in protein but not in DNA; phosphorus is present in DNA but not in protein)
   b. Upon infection of bacterial host cells with radiolabeled phage, only radioactive phosphorus entered the host cell and was passed on to the next generation
2. Because DNA containing the radiolabel phosphorus entered the host cell and was passed from one generation to the next, DNA must be the genetic material of phage T2

### D. RNA as genetic material

1. Some small viruses do not contain DNA, but only ***ribonucleic acid (RNA)*** and protein
   a. In 1957, Heinz Frankel-Conrat and B. Singer dissociated two strains of tobacco mosaic virus particles into their molecular components
   b. RNA from the first strain was mixed with protein from the second strain, and vice versa
   c. Infection of host tissue with the chimeric virus particles and examination of progeny virus revealed that the type of progeny strain always was dictated by the type of RNA in the chimeric particle
2. Thus, RNA serves as the genetic material in some viruses (such as the tobacco mosaic virus)

## II. Structure of Nucleic Acids

### A. General information

1. Nucleic acids (DNA and RNA) are polymers
2. The repeating unit of structure, or monomer, is the ***nucleotide***

### B. Chemical composition

1. A nucleotide is comprised of three parts: a sugar group, a nitrogenous base, and a phosphate group
   a. The sugar group consists of a 5-carbon (pentose) ring
      (1) DNA nucleotides contain deoxyribose (deoxyribonucleotides)
      (2) RNA nucleotides contain ribose (ribonucleotides)

(3) Sugar carbons commonly are numbered with a prime (′) symbol to differentiate them from the carbons that exist in nitrogenous bases

b. A nucleotide contains one of five different bases: adenine (A), thymine (T), cytosine (C), guanine (G), or uracil (U)

(1) Adenine and guanine are double-ringed bases called ***purines***

(2) Thymine, cytosine, and uracil are single-ringed bases called ***pyrimidines***

(3) Uracil does not occur in DNA

(4) In RNA, thymine is replaced by uracil

2. The combination of a sugar group and nitrogenous base (without a phosphate group) is called a ***nucleoside***

a. The adenine-containing nucleosides in DNA and RNA are deoxyadenosine and adenosine

b. The thymine-containing nucleoside in DNA is deoxythymidine

c. The cytosine-containing nucleosides in DNA and RNA are deoxycytidine and cytidine

d. The guanine-containing nucleosides in DNA and RNA are deoxyguanosine and guanosine

e. The uracil-containing nucleoside in RNA is uridine

3. A nucleotide's phosphate group is attached at the 5′ carbon of the pentose sugar

4. Nucleotides comprising DNA molecules occur in specific ratios, as determined by Erwin Chargaff in 1950 and stated in *Chargaff's rules* below

a. The number of purines equals the number of pyrimidines (A + G = C + T)

b. The number of adenine-containing nucleotides equals the number of thymine-containing nucleotides (A = T)

c. The number of cytosine-containing nucleotides equals the number of guanine-containing nucleotides (C = G)

**C. The DNA double helix**

1. One of the most significant biological discoveries ever was made by James Watson and Francis Crick in 1953 when they determined the arrangement of nucleotides in a DNA molecule

2. Watson and Crick relied on three lines of evidence for their model of DNA structure: Chargaff's rules, X-ray diffraction photography of DNA, and chemical component ratios of DNA

3. Nucleotides are connected to one another by phosphodiester linkages

a. The 5′ phosphate group of one nucleotide interacts with a 3′ hydroxyl group of another nucleotide

b. The resulting covalent bond results in a phosphate bridge between pentose sugars of adjoining nucleotides

c. The molecule demonstrates chemical polarity because the free 3′ carbon of each nucleotide is bonded with the phosphate group located on the 5′ carbon of the adjacent nucleotide; polarity is described as 5′ → 3′

d. A polynucleotide strand can be visualized as a sugar-phosphate backbone with nitrogenous bases projecting from each monomer

4. DNA polynucleotide strands interact with one another via the hydrogen bonds between nitrogenous bases to form double-stranded molecules (see *DNA Molecule,* page 38)

a. Because of the specificity of base pairing, polynucleotide strands comprising a double-stranded molecule are called complementary; if the nucleotides

## DNA Molecule

In DNA, polynucleotide strands interact with each other to form double-stranded molecules. The illustration below shows the arrangements of the adenine (A), cytosine (C), guanine (G), and thymine (T) nucleotides as well as the phosphate (P) and sugar groups in a eukaryotic DNA molecule; hydrogen bonds are indicated by dotted lines.

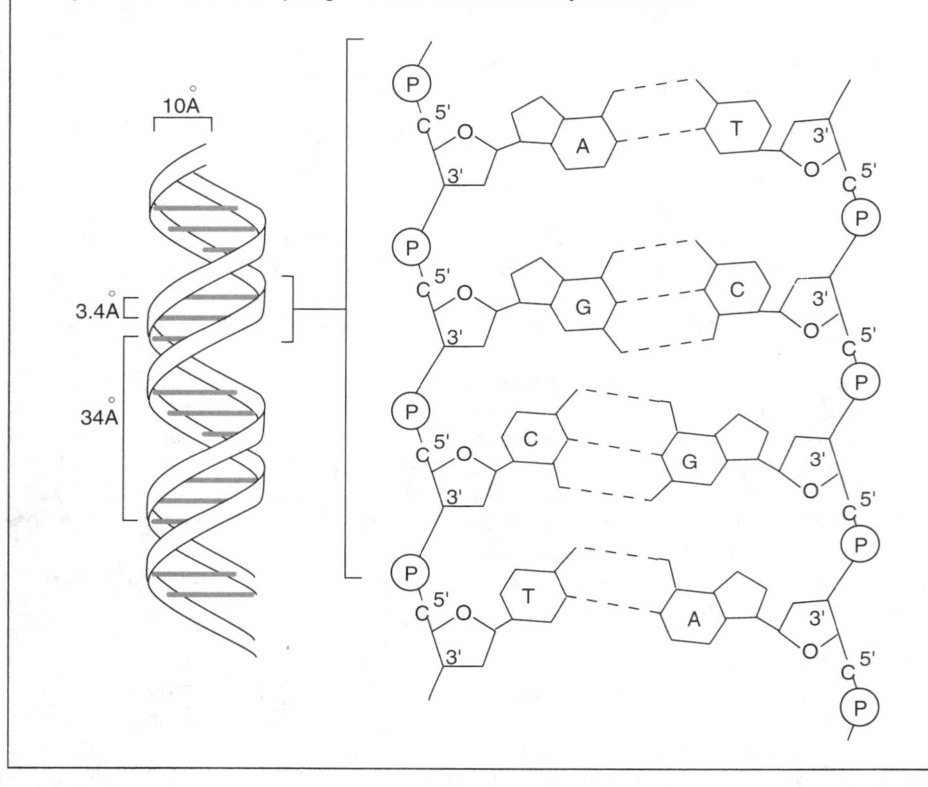


in one strand are known, the corresponding nucleotides in the other strand also are known

b. The purine adenine forms two hydrogen bonds with the pyrimidine thymine

c. The pyrimidine cytosine forms three hydrogen bonds with the purine guanine

d. Because G-C pairing (which comprises 3 hydrogen bonds) is stronger than A-T pairing (which comprises 2 hydrogen bonds), regions of DNA that have many G-C pairs are extremely stable

e. The complementary nature of the two strands makes DNA uniquely suited to store and transmit genetic information

5. The secondary structure of the double-stranded molecule resembles a helix, or a spiral staircase, with hydrogen-linked bases forming the individual steps
6. The two polynucleotide strands are oriented in an ***antiparallel*** arrangement; in other words, their backbones have opposite chemical polarities
7. Because DNA double helices are flexible, they can exist in a variety of structures, or conformations; this is called *conformational flexibility*

   a. These forms include the A-form, the B-form, and Z-DNA

   b. Forms differ in the number of base pairs per turn of the helix

c. The B-form, which is a right-handed helix, occurs in most physiological states and can shift into Z-DNA
d. The A-form, a minor variant, occurs in high concentrations of salts
e. Z-DNA is a left-handed zigzagged helix that can shift into the B-form
f. Gene function may be regulated, in part, by the structure of DNA

### D. Implications of DNA structure

1. The structure of DNA suggests that the order of base pairs may be related to gene function
2. The complementary nature of polynucleotide strands in a DNA double helix gave researchers immediate insight into the process of replication
3 The momentous discovery of Watson and Crick marked the beginning of a new era in genetic research that continues to reveal details about the storage and transfer of genetic information

## III. Semiconservative Replication of DNA

### A. General information

1. Scientists noticed that if complementary DNA strands are separated, each strand can serve as a template for the construction of new, complementary polynucleotide strands
2. The two resulting double helices each contain one polynucleotide strand from the original molecule and one new strand; this is called ***semiconservative replication***

### B. The Meselson-Stahl experiment

1. Data supporting the theory of semiconservative replication were provided by Matthew Meselson and Franklin Stahl in 1958
2. Equilibrium density-gradient centrifugation was used to distinguish between DNA strands that contained different isotopes of nitrogen
   a. *Escherichia coli* (a prokaryote) cells were grown in the presence of $^{15}_{7}N$ (the heavy isotope of nitrogen that has 8 neutrons and 7 protons) for many generations
   b. These cells were then transferred to a medium containing $^{14}_{7}N$ (the native form of nitrogen with 7 neutrons and 7 protons) for specific periods of time; afterward, the nitrogen content of the *E. coli* DNA was analyzed
   c. After one generation in medium containing $^{14}N$, the DNA had a hybrid density consisting of equal amounts of $^{14}N$ and $^{15}N$
   d. After two generations in medium containing $^{14}N$, half the DNA had a hybrid density and half contained only $^{14}N$
3. Patterns of nitrogen isotope incorporation during DNA replication were consistent with a semiconservative model, as predicted by Watson and Crick (see *Semiconservative DNA Replication,* page 40)

### C. Autoradiography of prokaryotic chromosomes

1. Semiconservative DNA replication can be visualized by growing bacterial cells in the presence of radioactive deoxyribonucleosides (such as $^{3}H$-labeled thymidine, which is radioactive)

## Semiconservative DNA Replication

Unwinding of the deoxyribonucleic acid (DNA) double helix produces two single-stranded templates for the synthesis of two new polynucleotide strands.

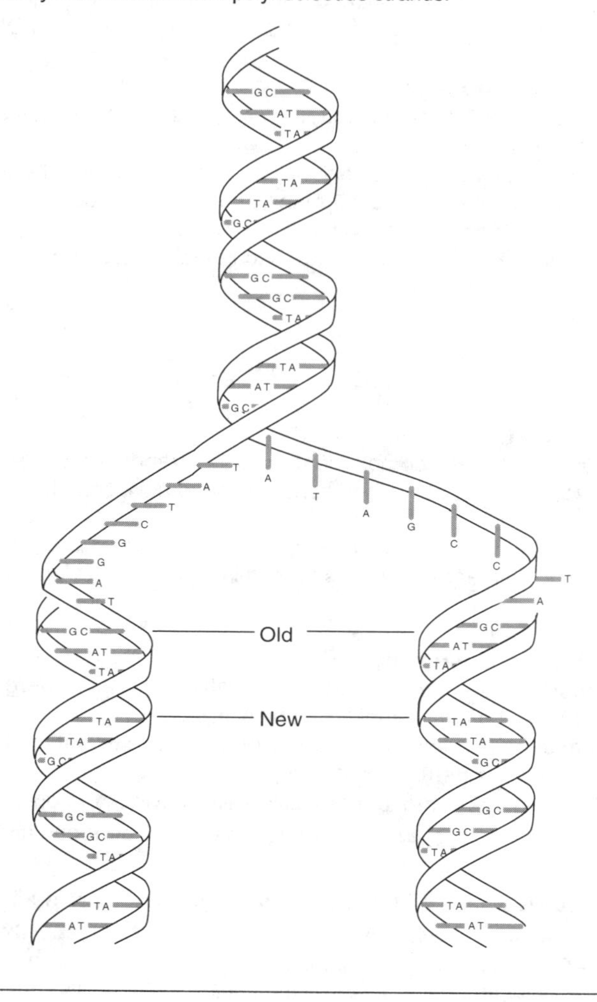


a. Thymidine is incorporated only into DNA

b. Cells in the process of DNA replication (and therefore $^3H$ incorporation) are lysed, their DNA collected gently, and the resulting preparations are coated with a photographic emulsion

c. Radioactive decay is recorded on film as the emulsion is exposed in the dark for a period of time

d. After the film is developed, scientists view the resulting photographs with a microscope

2. The separation of complementary DNA strands for the semiconservative synthesis of new polynucleotide strands can be visualized as a Y-shaped replication fork

3. Y-shaped forks originate at the point where replication begins, called the *origin*
   a. Prokaryotic chromosomes are circular and have one origin of replication and one terminus
   b. Chromosomes in the process of replication resemble the Greek letter theta (θ) and are called theta structures
   c. Replication begins at the origin and proceeds away from it bidirectionally

### D. Autoradiography of eukaryotic chromosomes

1. Eukaryotic DNA can be labeled with radioactive thymidine in a manner that is analogous to that used for proving the semiconservative replication of *E. coli*
2. After one round of DNA replication in the presence of radioactive thymidine, each daughter DNA molecule is composed of one radioactive polynucleotide strand and one nonradioactive polynucleotide strand
3. If eukaryotic cells are allowed to undergo an additional round of DNA replication in the absence of radioactive thymidine, the resulting cells will contain two types of DNA double helices; one type will contain both radioactive and nonradioactive polynucleotide strands and the other will contain two nonradioactive strands
   a. The two types of DNA helices represent sister chromatids
   b. One chromatid will be radioactive and the other will not be radioactive
4. These findings indicate that replication of DNA in eukaryotic chromosomes, as in prokaryotic chromosomes, is semiconservative

## IV. Mechanisms of DNA Replication

### A. General information

1. Although the concept of semiconservative replication is elegant in its simplicity, the mechanisms underlying DNA synthesis are complex
2. Replication requires separation of polynucleotide strands (breaking of hydrogen bonds), stabilization of the resulting single-stranded DNA, initiation of DNA synthesis, polymerization of two new polynucleotide strands, proofreading of the nascent strands to ensure fidelity, and twisting of the replicating DNA molecule to alleviate torsion created during replication
3. The replication apparatus, or *replisome,* is an elaborate multienzyme complex
   a. For example, *E. coli* requires at least two dozen proteins for DNA replication
   b. The required enzymes and proteins together make up the replisome
4. Replication of DNA occurs during the S phase of the eukaryotic cell cycle

### B. The replication fork

1. Most of our knowledge of the events that occur at the replication fork is derived from in vitro studies of *E. coli*
2. Replication begins at a specific chromosomal region called the origin
   a. The *E. coli* chromosome has a single origin
   b. Eukaryotic chromosomes have multiple origins and, hence, multiple replicating units or *replicons*
3. Base pairing at the origin is disrupted, leading to localized denaturation and unwinding of the double helix
   a. The enzyme required for denaturation and unwinding is called ***helicase***
   b. *E. coli* contains several helicases, one of which is the product of the *rep* gene

c. Adenosine triphosphate (ATP) hydrolysis is required for helicase function
4. In most organisms, unwinding of DNA continues away from the origin in two directions, resulting in bidirectional replication
5. Unwinding of the helix by helicase results in the formation of single-stranded templates
   a. Single-strand DNA-binding (SSB) proteins stabilize the unwound region of the replication fork
   b. In the absence of SSB proteins, single-stranded regions would form hydrogen bonds with the complementary strand or the complementary region in the same strand
6. Unwinding of the DNA at the replication fork results in overwinding of the DNA elsewhere on the molecule
   a. Overwinding is called positive ***supercoiling,*** while unwinding is called negative supercoiling
   b. Negative supercoiling is required for the replication of DNA molecules
   c. In order for the replication to proceed, positive supercoiling must be enzymatically removed by gyrases
      (1) Gyrases remove positive supercoils as they accumulate ahead of the moving fork
      (2) Gyrases belong to a class of enzymes called ***topoisomerases,*** which alter DNA topology by inducing positive or negative supercoils ahead of the replication fork in preparation for replication
7. DNA that is unwound and stabilized is ready to serve as a template for the synthesis of a complementary strand (see *The Replication Fork*)

**C. DNA polymerases**

1. ***DNA polymerases*** are complex enzymes which participate in a variety of reactions, including DNA replication and repair of DNA damage
   a. *E. coli* contains three DNA polymerases: pol I, pol II, and pol III
   b. Eukaryotes contain at least four DNA polymerases: $\alpha$, $\beta$, $\gamma$, and $\delta$
   c. The specific roles of all identified polymerases have not been determined yet
2. DNA polymerase is responsible for catalyzing the covalent addition of deoxyribonucleotides onto an existing polynucleotide strand
   a. DNA polymerase requires magnesium ions ($Mg^{+2}$) and the triphosphate form of the four deoxyribonucleotides (that is, deoxyadenosine 5′ triphosphate [dATP])
   b. DNA polymerase requires a primer and a template
      (1) The enzyme ***primase,*** along with other protein components of the primosome (which, in turn, is part of the replisome), binds to single-stranded DNA and produces a short RNA primer, thus providing a free 3′-OH as a starting point for DNA synthesis
      (2) DNA polymerase also requires a preexisting strand of DNA to act as a template to direct the incorporation of the nucleotides that contain complementary bases
   c. A phosphodiester linkage is formed between the 5′-phosphate of the incoming deoxyribonucleotide triphosphate and the 3′-OH of the previously incorporated nucleotide
   d. The direction of polymerization is $5' \rightarrow 3'$
3. Specific DNA polymerases (such as *E. coli* DNA polymerases I and III and eukaryotic DNA polymerase $\gamma$) are responsible for proofreading the nascent

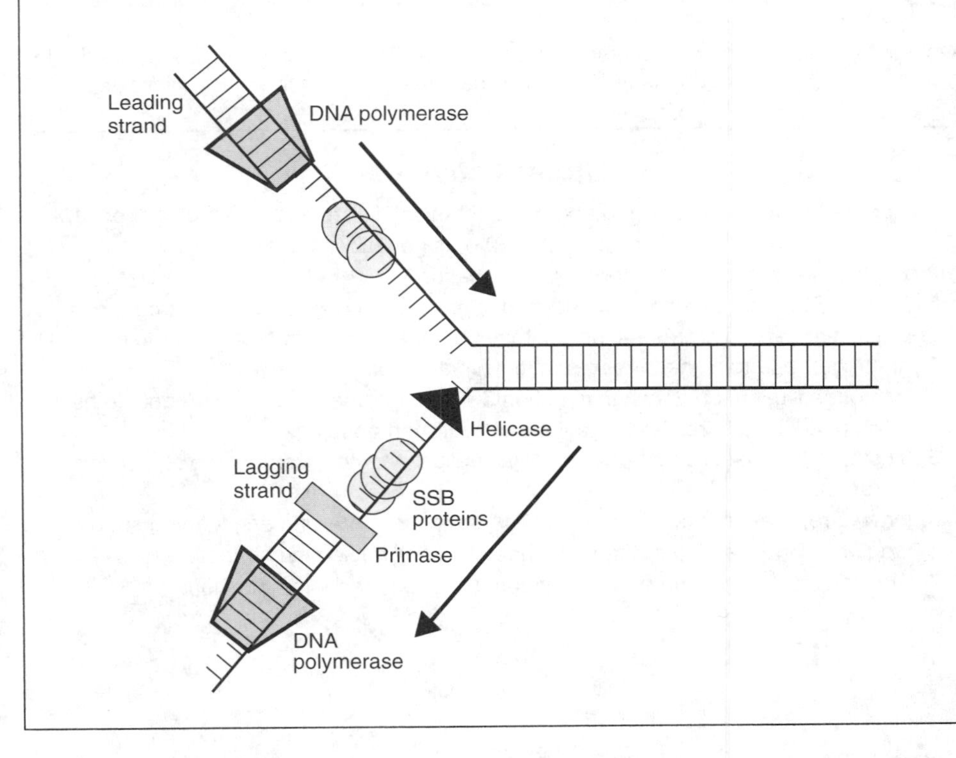


**The Replication Fork**

The diagram below illustrates a DNA replication fork. Because of the opposite polarities of the polynucleotide strands in a DNA double helix, one new strand is synthesized toward the site of replication (called the leading strand) and the other new strand is synthesized away from the fork (called the lagging strand).

DNA strand and removing mismatched nucleotides via 3′ → 5′ ***exonuclease*** activity

a. An exonuclease degrades DNA by sequential removal of nucleotides from the end of the polynucleotide strand

b. This activity ensures the fidelity of DNA replication

4. Other DNA polymerases (*E. coli* DNA polymerase I, for example) are responsible for removal of the RNA primers via 5′ → 3′ exonuclease activity

**D. Continuous vs. discontinuous DNA synthesis**

1. The antiparallel nature of the DNA double helix and the 5′ → 3′ direction of DNA polymerization dictate that the two nascent strands at the replication fork will move in opposite directions
   a. One strand, the leading strand, will be synthesized towards the fork
   b. The other strand, the lagging strand, will be synthesized away from the fork
2. The leading strand, synthesized in the same direction as the moving replication fork, is synthesized continuously

3. Because the lagging strand is synthesized in the opposite direction of the moving replication fork, its synthesis must be reinitiated periodically in a newly exposed region of the template; it is synthesized discontinuously
   a. Reinitiation requires the synthesis of additional RNA primers
   b. In 1968, Reiji Okazaki and others noted that discontinuous DNA synthesis results in the formation of many small DNA fragments, now called ***Okazaki fragments***
   c. These fragments are sealed together by ***DNA ligase,*** which catalyzes the formation of phosphodiester linkages between adjacent nucleotides

---

## Study Activities

1. Apply what you have learned about the chemical nature and arrangement of DNA strands to explain why nucleic acids serve as genetic material.
2. Explain the concept of bacterial transformation.
3. Draw the structure of a nucleotide by using a circle to represent the nitrogenous base, a square to represent the pentose sugar, a triangle to represent the phosphate group, and a line to represent covalent bonds.
4. Using the illustration drawn in question 3 as a monomer, draw a DNA double helix. Use dotted lines to represent regions of hydrogen bonding.
5. Illustrate the relationship between DNA replication and chromosome replication in eukaryotes.
6. Draw a replication fork. Include the DNA template, RNA primers, leading strand, and lagging strand. Label the 5′ and 3′ ends of all polynucleotide strands.
7. Summarize the enzyme and protein functions required for replication.

# 6

# Structure of the Eukaryotic Chromosome

## Objectives

After studying this chapter, the reader should be able to:

- Describe the relationships that exist among deoxyribonucleic acid (DNA), histones, and chromatin.
- Illustrate the structure of a nucleosome.
- Describe the levels of DNA packaging.
- Differentiate between euchromatin and heterochromatin.
- Explain the basis for mosaicism in certain heterozygous female mammals.
- Describe the possible arrangements of DNA sequences within a genome and the frequencies at which they occur.
- Discuss the function and structure of centromeres and telomeres.
- Describe how differences in chromosome staining and morphology can be explained by the chemical composition and genetic state of specific chromosomal regions.

## I. Size and Complexity of Eukaryotic Genomes

### A. General information

1. Eukaryotic cells range from 10 to 100 μm in size; they are approximately 10 times the size of prokaryotic cells (which are 1 to 10 μm)
2. A human cell, however, contains a *genome* (total complement of genetic material) that is approximately 1,000 times as large as that of an *Escherichia coli* cell
   a. Each human cell contains approximately 6 billion base pairs (bp) of DNA
   b. If the DNA base pairs of one human cell were strung end-to-end, it would be about 2 m long
   c. Eukaryotic DNA must therefore be packaged into chromosomes very efficiently

### B. One DNA molecule per chromosome

1. Various experimental data support the hypothesis that every eukaryotic chromosome contains a single, unbroken, double-stranded DNA molecule
   a. Autoradiography of chromosomes (described in Chapter 5, Structure and Replication of Genetic Material) provided visual evidence that each chromatid contains a single DNA molecule
   b. Measurement of DNA elasticity in solution supplied estimates of the length of individual DNA strands

(1) DNA molecules in solution were stretched by a rotating paddle and then allowed to recoil; the recoil time is proportional to the length of the longest molecule

(2) When chromosome size was experimentally altered, the resulting recoil times also changed; this indicates that a chromosome was composed of a single DNA molecule

c. Electron microscopy of DNA preparations allowed direct measurement of DNA molecule length

d. Advanced electrophoresis (discussed in Chapter 10, The Unit of Function: Molecular Analysis) facilitated the physical separation of individual, unbroken DNA molecules

(1) Electrophoresis separates DNA molecules according to size

(2) Molecules are deposited into gelatinous material and then subjected to an electric field

(3) Because DNA molecules are negatively charged, they migrate through the gel toward the positive electrode at a rate that is proportional to their size

2. The DNA in one human cell therefore represents 46 individual DNA molecules that are packaged into 46 chromosomes

## II. DNA Packaging

### A. General information

1. DNA packaging refers to the compaction of a large DNA molecule into a relatively small area to form a single chromosome
2. Eukaryotic chromosomes are composed of a complex of DNA, ribonucleic acid (RNA), and protein called ***chromatin***

### B. Chromatin structure

1. Two types of proteins exist in chromatin: ***histones,*** which are basic (positively charged) proteins, and *non-histones,* which are acidic (negatively charged) proteins
2. The amount of histones in chromatin is roughly equal to the amount of DNA
3. Histones that have been isolated from different organisms are remarkably similar; this indicates that they perform the same function in all organisms
4. All histones contain varying amounts of the basic amino acids lysine and arginine; this accounts for their overall positive charge
5. Histones can be divided into five main types according to their amino acid composition; they are H1, H2A, H2B, H3, and H4
6. Histones H2A, H2B, H3, and H4 are part of a structure called the nucleosome core particle
   a. Each histone in a nucleosome core particle is represented by two molecules
   b. Because it contains eight molecules, the nucleosome core particle is called a histone octamer

### C. Nucleosome structure

1. The first level of DNA packaging in the eukaryotic chromosome involves the periodic wrapping of a DNA molecule — which is 2 nm wide and contains approximately 200 bp — around a nucleosome core particle

2. The nucleosome core particle and the associated DNA together form the ***nucleosome***
   a. Each nucleosome is approximately 11 nm in diameter and 5.7 nm thick and has an ellipsoidal shape
   b. 146 bp of the DNA is wrapped one-and-three-fourths of the way around each nucleosome core particle
   c. Approximately 54 bp of DNA exist between one nucleosome core particle and another; these DNA bp are called *linker DNA*
   d. Together, nucleosomes resemble beads on a string

**D. Higher-order packaging**

1. The next level of packaging involves the further coiling of the nucleosomes into a supercoiled structure
   a. This structure is called a *solenoid* because of its resemblance to the coiled wire that is part of an electromagnet
   b. The solenoid structure is about 30 nm thick
   c. The histone H1 stabilizes the structure (see *Nucleosome and Solenoid Structure,* page 48)
2. A third level of packaging involves the looping and attachment of the 30-nm solenoid thread to a non-histone scaffold
3. The scaffold assumes the shape of a metaphase chromosome
4. The fully condensed metaphase chromosome is approximately 1,400 nm wide

## III. Differentiation Within a Chromosome

**A. Types of chromatin**

1. Chromatin exists in two forms: ***euchromatin*** and ***heterochromatin***
   a. Euchromatin is light-staining chromatin that condenses only during nuclear division
   b. Heterochromatin is dark-staining chromatin that becomes highly condensed during mitosis and other phases of the cell cycle
   c. Chromatin stains include carmine, orcein, toluidine blue, and methylene blue
2. Euchromatin is genetically active, whereas heterochromatin is genetically inert
3. Heterochromatin may represent permanently condensed regions (*constitutive* heterochromatin) or regions that were previously euchromatic (*facultative* heterochromatin)
   a. Heterochromatin that exists at the centromere and ***telomere*** are examples of constitutive heterochromatin
   b. An example of facultative heterochromatin in female mammals is the ***Barr body,*** which contains the inactive X chromosome

**B. Inactivation of the X chromosome**

1. Mammalian female chromosomes (XX) contain two copies of all X-linked genes, whereas male chromosomes (XY) contain only one copy
2. To compensate for the difference in gene dosage, inactivation of one X chromosome occurs in female somatic cells
   a. The inactive, heterochromatic chromosome is called a Barr body
   b. Inactivation is random and occurs during embryo development
   c. This results in one genetically active X chromosome per cell

## Nucleosome and Solenoid Structure

This illustration of the nucleosome and solenoid structure shows the location of histone H1 relative to the nucleosome.

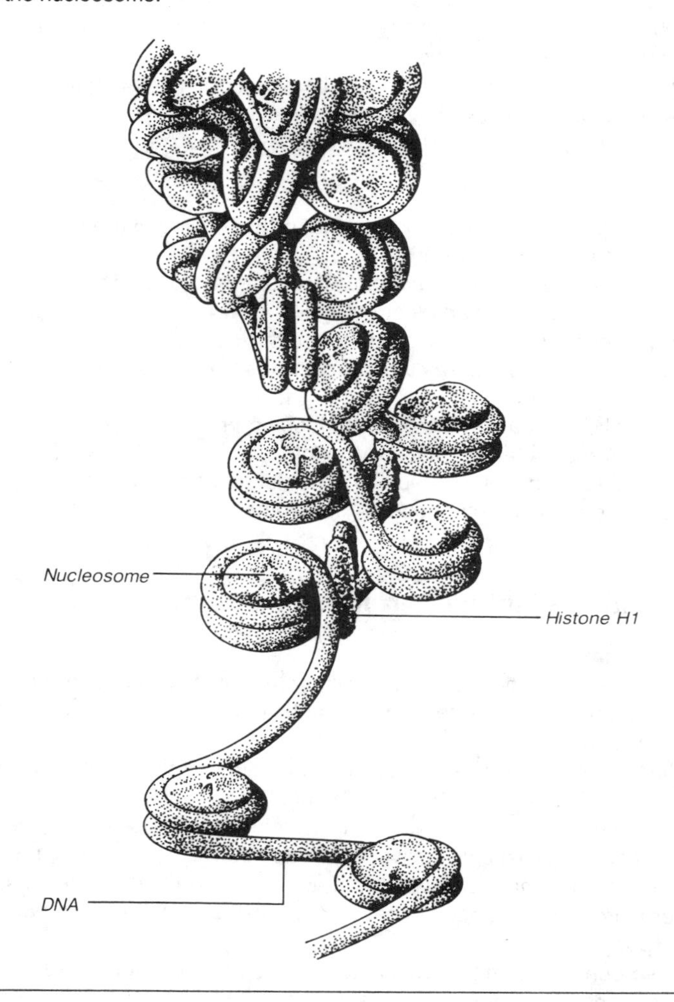


3. Some X-linked traits give rise to ***mosaicism*** — the possession of genetically different cell types — in individuals that are heterozygous for such a gene
   a. For example, the random occurrence of orange and black fur on calico and tortoiseshell cats results from the random inactivation of the X chromosome that contains the orange allele *(O)* or the black allele *(o)* in heterozygous *(Oo)* female cats
   b. Human females who are heterozygous for anhidrotic ectodermal dysplasia exhibit mosaicism in sweat glands
4. X-chromosome inactivation is one form of gene dosage compensation

# IV. DNA Sequence Organization

## A. General information

1. Double-stranded DNA molecules are held together by hydrogen bonds between the nitrogenous bases
2. Heating (boiling) a DNA solution breaks the hydrogen bonds and liberates the two single-stranded DNA molecules
   a. Disruption of hydrogen bonding in DNA molecules is called *denaturation*
   b. Subsequent cooling results in renaturation or ***reannealing*** of complementary DNA sequences
3. The kinetics of renaturation depend upon the complexity of the reannealing DNA sequences
   a. DNA sequences that are represented by multiple copies will renature the most rapidly
   b. Single-copy DNA sequences will renature the most slowly
4. By studying renaturation kinetics, geneticists have discovered that various classes of DNA sequences exist
   a. Unique sequences generally occur once in a haploid genome
   b. Moderately repetitive sequences occur, on average, in 1,000 to 100,000 copies per haploid genome
   c. Highly repetitive sequences typically occur in approximately 1,000,000 copies per haploid genome

## B. Repetitive DNA and gene families

1. The moderately repetitive class of DNA sequences includes those encoding histones, insect eggshell proteins, actin and myosin (cytoplasmic proteins), hemoglobins, ribosomal RNA, and ***transfer RNA*** (which are important in protein synthesis) as well as other gene sequences with no known function
   a. Functional repetitive sequences are called ***gene families***
      (1) Members of a gene family may be slightly dissimilar at the nucleotide level but remain highly homologous and encode products of similar function
      (2) Members of a gene family that are rendered nonfunctional by mutation are called ***pseudogenes***
      (3) Individual, repeated genes may be arranged in clusters or may be dispersed throughout the genome
         (a) Clustered, repeated DNA sequences are arranged in uninterrupted tandem arrays of repeated units (for example, ribosomal RNA genes)
         (b) Interspersed, repeated DNA sequences are dispersed among unique sequences (for example, histone genes)
   b. Some repeated DNA sequences have poorly characterized or no known functions
      (1) Short interspersed elements (SINEs) have repeating units that consist of less than 500 bp
      (2) Long interspersed elements (LINEs) have repeating units that consist of several thousand bp
      (3) All eukaryotic organisms contain SINEs and LINEs
2. Highly repetitive DNA sequences range in size from several base pairs to several thousand base pairs

a. These sequences do not code for a functional product
b. Many of these sequences contain a higher percentage of guanine and cytosine than the majority of DNA sequences, and are called satellite DNA because of their different density
c. Highly repetitive DNA sequences are associated with heterochromatin, and therefore occur near centromeres and telomeres

### C. Centromere and telomere sequences

1. Centromeres are involved in the movement of chromosomes during mitosis and meiosis
   a. The spindle apparatus, which consists of microtubules and microfilaments, attaches to the centromere at the ***kinetochore***
      (1) In yeast cells, for example, only one microtubule is attached to each centromere during mitosis and meiosis
      (2) In more complex eukaryotes, several microtubules are associated with each centromere during mitosis and meiosis
   b. Centromeres have specific nucleotide sequences
      (1) The centromeres of yeast cells are slightly different, but each one contains approximately 220 bp and resembles the others
      (2) Different organisms contain different centromeric sequences
2. Telomeres are chromosome termini
   a. Telomeres stabilize broken, linear chromosomes, which stick to other chromosomes readily and may lead to chromosomal abnormalities
   b. Telomeres in a given species have the same nucleotide sequence
   c. Telomeric nucleotide sequences may be divided into two types: telomere-associated sequences and simple telomeric sequences
      (1) Telomere-associated sequences are complex, repeated sequences that extend inward from the end of a chromosome
      (2) Simple telomeric sequences are short sequences that are repeated in tandem at the terminus of a chromosome; in humans, this sequence is 5′-TTAGGG-3′
   d. In some cases, telomeric sequences assume a secondary structure; one strand of the DNA double helix extends beyond the other and bends back on top of itself to form a hairpin structure at the terminus of a chromosome
      (1) These secondary structures are stabilized by hydrogen bonds
      (2) The secondary structure of the telomere is essential for normal chromosome replication

## V. Specialized Chromatin Structures

### A. General information

1. The chemical composition and the genetic state (that is, active or inactive) of specific chromosomal regions can produce visible differences in staining and chromosome morphology
2. These visible structures serve as landmarks used to identify specific chromosomes and help geneticists elucidate chromosome structure

**B. Chromosome banding**

1. Staining of mitotic chromosomes with the Giemsa stain produces patterns of light-staining (G-light) and dark-staining (G-dark) transverse bands
   a. The Giemsa stain is a mixture of dyes based on azure and eosin
   b. DNA that is rich in guanine and cytosine (GC-rich) stains lightly; DNA that is rich in adenine and thymine (AT-rich) stains darkly
   c. Staining with quinacrine mustard results in fluorescent Q bands that also reveal AT-rich regions
2. The chromosomes of the *Drosophila* (fruit fly) salivary gland have transverse banding patterns that result from differential chromatin condensation
   a. *Drosophila* salivary gland chromosomes are unique structures because they undergo many rounds of chromosome replication in the absence of karyokinesis
   b. Each giant, multi-stranded chromosome is called a ***polytene chromosome*** and contains approximately 1,024 DNA molecules
   c. Studies performed by Arthur Chovnick in 1979 demonstrated a correlation between the number of chromosome bands and the number of genes
   d. Recent studies, however, have not yielded consistent data; the data show that genes (about 17,000) may outnumber chromosome bands (about 6,000)

**C. Chromosome puffs**

1. The salivary gland chromosomes of *Drosophila* also exhibit diffuse, swollen areas called *chromosome puffs* or *Balbiani rings*
2. Chromosome puffs are areas of chromosome uncoiling, a process that facilitates gene activity
3. They are categorized as stage-specific (displayed only in certain stages of development), tissue-specific (displayed in certain tissues), environmentally induced (displayed in certain environments), or constitutive (displayed all the time)

**D. Lampbrush chromosomes**

1. Areas of chromatin uncoiling and associated gene activity have been visualized in amphibian oocytes
2. In these oocytes, loops of chromatin originate from a central axis, resembling the shapes of the brushes once used to clean oil lamps; accordingly, these chromosomes are called lampbrush chromosomes

---

## Study Activities

1. Explain why eukaryotic DNA sequences need to be packaged into chromosomes.
2. Draw a nucleosome. Label the core particle and the DNA strand, and indicate the location of histones H1, H2A, H2B, H3, and H4.
3. Define euchromatin, facultative heterochromatin, and constitutive heterochromatin.
4. If an orange male cat of genotype $X^{O}Y$ mates with a calico female of genotype $X^{O}X^{o}$, what are the possible genotypes and phenotypes of the progeny?
5. Make a table that summarizes the different classes of DNA sequence organization. Include characteristic features and examples of each type of sequence.
6. Describe the structural basis of chromosome bands and puffs.

# 7

# Linkage, Recombination, and Gene Mapping

## Objectives

After studying this chapter, the reader should be able to:

- Differentiate between linkage, crossing-over, and recombination.
- Calculate frequencies of recombination for linked loci.
- Construct genetic maps using information derived from three-factor testcrosses.
- Describe the phenomenon of interference.
- Explain the usefulness of mapping gene functions.
- Determine the distance between loci in haploid ascus-forming organisms.
- Calculate the distance between a gene and the centromere in haploid ascus-forming organisms.
- Describe the usefulness of human-rodent hybrid cell lines for gene mapping in humans.

## I. Linkage

### A. General information

1. Mendel's law of independent gene assortment is a direct reflection of chromosome behavior during meiosis; the separation of one pair of homologs is independent of the separation of all other homologous pairs
2. Mendel, however, did not consider ***linkage*** — when genes reside on the same chromosome
3. Genes that are physically located on the same chromosome are called *linked genes*
4. Linked genes tend to segregate together, thereby causing deviations from the ratios predicted by Mendel's law of independent assortment
5. In order for linked genes to recombine (that is, become associated with different alleles), chromosomes must undergo a physical exchange of genetic material during meiosis

### B. The discovery of linkage

1. William Bateson, E.R. Saunders, and Reginald Punnett were the first geneticists to recognize a deviation from predicted segregation ratios among $F_2$ progeny in their experiments with pea plants in 1905
   a. Pea plants having purple flowers and long pollen were crossed with pea plants having red flowers and round pollen

b. The resulting $F_1$ progeny had purple flowers and long pollen; the purple flower allele *(P)* and the long pollen allele *(L)* were dominant to the red flower allele *(p)* and the round pollen allele *(l)*, respectively

c. $F_2$ progeny were expected to occur in a characteristic dihybrid 9:3:3:1 ratio — with 56.25% having purple flowers and long pollen, 18.75% having purple flowers and round pollen, 18.75% having red flowers and long pollen, and 6.25% having red flowers and round pollen

d. However, the resulting $F_2$ progeny consisted of the following phenotypes: 74.6% having purple flowers and long pollen, 5.5% having purple flowers and round pollen, 5.5% having red flowers and long pollen, and 14.4% having red flowers and round pollen

2. Bateson, Saunders, and Punnett recognized that the parental gene combinations (purple flower with long pollen and red flower with round pollen) were represented in the $F_2$ generation at frequencies higher than expected

   a. The higher frequencies of these two phenotypes presumably resulted from a higher percentage of *PL*-containing and *pl*-containing pollen

   b. The two dominant alleles therefore appeared to show coupling, as did the two recessive alleles

3. The explanation for the observed coupling of alleles came from the laboratory of T.H. Morgan in 1911

   a. Morgan noted that two X-linked alleles — *white eye* and *miniature wing* — in *Drosophila* showed coupling

   b. He concluded that coupling occurred because both genes resided on the same chromosome

   c. He also concluded that the coupled genes sometimes would recombine, thus giving rise to ***recombinant*** phenotypes; this is called ***recombination***

   d. Morgan hypothesized that an exchange of genetic material occurred between the two X chromosomes during some meioses

   e. The process by which homologous chromosomes exchange material during meiosis is called ***crossing-over*** (see *Crossing-over*, page 55)

## C. Revisiting the dihybrid testcross

1. Consider a hypothetical dihybrid cross between an individual homozygous for linked genes *A* and *B* and an individual homozygous for linked genes *a* and *b*

   a. In order to indicate the linkage for this particular testcross, the genotypes of these two individuals are designated as *AB/AB* and *ab/ab*

   b. This nomenclature tells us that genes listed on one side of the slash are linked on one homolog, while the alleles on the other side of the slash are linked on the other homolog

2. The resulting $F_1$ generation can be designated as *AB/ab* (note that linked genes are grouped on the same side of the slanted line)

   a. In the absence of crossing-over (complete linkage), the $F_1$ generation produces two types of gametes in equal proportions: *AB* and *ab*

      (1) Both of these gametes contain allele combinations found in the parents, which also is called the parental combination of alleles

      (2) If an $F_1$ individual is crossed with an *ab/ab* individual, only two types of progeny will result in equal proportions: *AB/ab* and *ab/ab*

      (3) Again, both types of testcross individuals will express the parental combinations of alleles

(4) The phenotypes of the testcross progeny therefore reflect the genotypes of the gametes produced by the $F_1$ generation

b. In the presence of crossing-over (partial linkage), the $F_1$ generation produces four types of gametes: *AB, ab, Ab,* and *aB*

(1) Two of these gametes contain parental allele combinations (*AB* and *ab*) and two contain recombinant allele combinations (*Ab* and *aB*)

(a) Because of partial linkage, however, the gametes will not occur in equal proportions

(b) The frequency of recombinant gametes will be somewhere between that expected for total linkage (0%) and that expected for independent assortment (50%)

(c) The frequency of recombinant gametes depends on the frequency of crossing-over

(d) For the sake of argument, let us assume that crossing-over occurs between the *A* and *B* genes during 10% of meioses; thus, 10% of the gametes will be recombinant (5% *Ab* and 5% *aB*) and 90% of the gametes will be parental (45% *AB* and 45% *ab*)

(2) If the $F_1$ progeny exhibiting crossover is crossed with an *ab/ab* individual, four types of progeny will result: *AB/ab, ab/ab, Ab/ab,* and *aB/ab*

(3) Two types of these testcross individuals are parental (*AB/ab* and *ab/ab*) and two types are recombinant (*Ab/ab* and *aB/ab*)

(4) The proportions of testcross progeny are the same as the proportions of gametes produced by the $F_1$ generation: 45% *AB/ab,* 45% *ab/ab,* 5% *Ab/ab,* and 5% *aB/ab*

c. Thus, the percentage of recombinant phenotypes (also called % recombination) observed in a dihybrid testcross is proportional to the percentage of crossing-over

## II. Recombination and Crossing-over

### A. General information

1. Because recombination requires the physical exchange of genetic material, some geneticists reasoned that this process may be visible in meiotic chromosomes
2. ***Chiasmata*** are cross-shaped structures that form between non-sister chromatids during the diplonema stage of meiosis I
3. Studies revealed a direct correlation between the frequency of the chiasmata and the frequency of crossing-over
4. In 1931, Harriet Creighton and Barbara McClintock proved that recombination was accompanied by a physical exchange of chromosomal segments

a. Creighton and McClintock produced maize plants that were heterozygous for linked genes and heterozygous for a visible chromosomal abnormality (one chromosome was longer because an extra DNA segment was added to it)

b. Crossing-over between the two genes always was accompanied by the exchange of the abnormal chromosome segment from one homologous chromosome to another

c. This proved that the genes were residing on the chromosome that was heterozygous for the visible abnormality and that recombination involved the breaking and rejoining of homologous chromosome segments

**Crossing-over**

This illustration compares the meiotic products resulting from the absence of a crossing-over with those resulting from a single crossing-over.

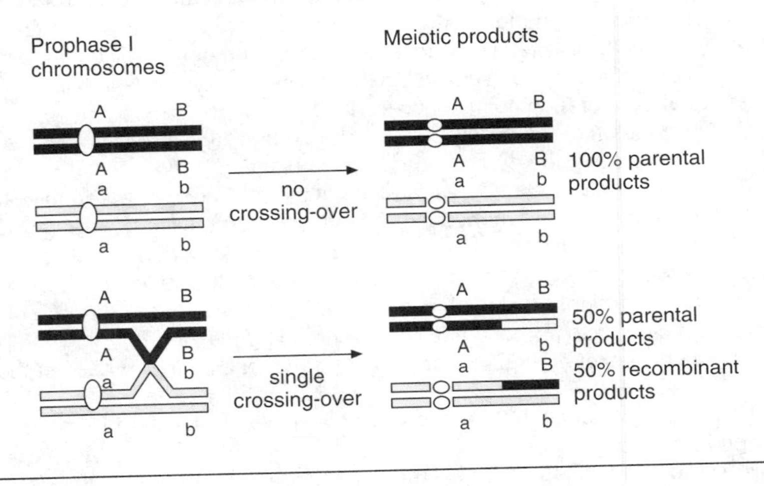


**B. Crossing-over in tetrads**

1. Crossing-over occurs during prophase I of meiosis and involves duplicated, synapsed homologs — the tetrads
2. Crossing-over can occur between any two non-sister chromatids
3. Because a single crossing-over affects only two of the four tetrad chromatids, 50% of the meiotic products are recombinant and 50% are parental
4. Even if a single crossing-over were to occur in every tetrad, only 50% of the products would be recombinant
5. Although a single crossing-over involves only two chromatids, multiple crossing-overs can occur between two genes
   a. If two crossing-overs occur between heterozygous genes and involve the same two chromatids (called a two-strand double crossover), then all meiotic products will be parental
   b. If two crossing-overs occur between heterozygous genes and the second crossing-over involves one of the chromatids that was involved in the first crossing-over (called a three-strand double crossover), then 50% of the products will be recombinant
   c. If two crossing-overs occur between heterozygous genes and the second crossing-over involves the two chromatids that were not involved in the first crossing-over (called a four-strand double crossover), then all products will be recombinant
   d. Because the chromatids involved in any one crossover event are random, two-strand, three-strand, and four-strand double crossovers occur in a 1:2:1 ratio; the average percentage of recombinant products is 50%
   e. Even if double crossover events were to occur in every tetrad, only 50% of the products would be recombinant
6. Thus, the percentage of recombination observed between any two heterozygous genes cannot exceed 50%

## III. Gene Mapping

### A. General information

1. The frequency of crossing-over between two genes, and therefore the percentage of recombination, depends upon the distance that separates the genes
   a. The probability that a crossover event between genes will occur is minimal when the distance separating the genes is very small (in this case, the percentage of recombination is low)
   b. For genes that are widely spaced, the probability of crossing-over is greater; in this case, the percentage of recombination is high
2. In 1913, Alfred Sturtevant reasoned that if the distances between three or more linked genes are determined experimentally, then the linear arrangement of these genes on the chromosome also can be determined
   a. The location of a gene on a chromosome is called its ***locus***
   b. The determination of the locus is called ***gene mapping***
   c. Homologous chromosomes contain homologous loci
   d. Two alleles of a gene are located at corresponding loci in each of the homologs
3. A collection of genes that are located on a single chromosome is called a *linkage group*
4. The number of linkage groups found in an organism is equal to the haploid number of chromosomes

### B. Mapping gene distance

1. The units of measure used for gene mapping are the ***map unit*** (m.u.), which more commonly is used, or the centimorgan (cM)
2. One m.u. is defined as the distance between loci for which the frequency of recombination is 1% (that is, 1 out of 100 meiotic products is recombinant)
   a. A map distance of 10 m.u. between genes *A* and *B* means that 1 out of every 10 meiotic products will be recombinant and that 9 out of 10 will be parental in this region
   b. For example, if individuals heterozygous for genes *B* and *C* produce recombinant gametes at a frequency of 1 in 20, then the distance between the *B* and *C* loci is 5 m.u.
3. Map units are additive; if genes *A, B,* and *C* are linked in the stated order and the distance between *A* and *B* and the distance between *B* and *C* are 3 m.u. and 2 m.u., respectively, then the distance between *A* and *C* is 5 m.u.

### C. The three-factor testcross

1. Consider the testcross of a homozygous ***wild-type*** *Drosophila* male *(Rs Cu E/rs cu e)* and a female homozygous for three linked, ***mutant*** genes: rosy eyes *(rsrs),* curly wings *(cucu),* and ebony body *(ee)*
   a. The resulting $F_1$ progeny is heterozygous at all three loci (Rs Cu E/rs cu e)
   b. When a female $F_1$ individual is crossed with an rs cu e/rs cu e male, the following progeny result: 66 wild-type flies (Rs Cu E/rs cu e); 19 ebony flies (Rs Cu e/rs cu e); 2 curly-winged flies *(Rs cu E/rs cu e)*; 14 rosy-eyed flies *(rs Cu E/rs cu e)*; 12 curly-winged, ebony flies *(Rs cu e/rs cu e)*; 2 rosy-eyed, ebony flies *(rs Cu e/rs cu e)*; 17 rosy-eyed, curly-winged flies *(rs cu E/rs cu e)*; and 68 rosy-eyed, curly-winged, ebony flies *(rs cu e/rs cu e)*

c. Note that the $F_1$ fly used to make the testcross was female; crossing-over does not occur in *Drosophila* males

d. Of the progeny given above, flies with the parental combination of alleles — *Rs Cu E/rs cu e* and *rs cu e/rs cu e* — are produced in the greatest number, whereas those resulting from double crossing-overs — *rs Cu e/rs cu e* and *Rs cu E/rs cu e* — are produced in the lowest number

2. Comparison of the parental genotypes with the genotypes of the double crossover progeny tells us the linear order of the genes
   a. Two of the alleles will maintain their linkage relationships after a double crossover, whereas the allele in the center will not
   b. Using the example above, *Rs* and *E* are still coupled in the double crossover progeny, as are *rs* and *e;* therefore, the *cu* gene is located between the other two genes
3. To determine the distance between the genes, we must consider each gene pair separately
   a. To determine the distance between the rosy eye gene and the curly wing gene, we need to calculate the percentage of recombinants
      (1) The flies that exhibit the parental phenotypes are wild-type at both loci or mutant at both loci (66 + 19 + 17 + 68 = 170 flies)
      (2) The flies that exhibit the recombinant phenotypes are mutant at one locus and wild-type at the other (2 + 14 + 12 + 2 = 30 flies)
      (3) The percentage of recombination equals the number of recombinants divided by the total number of flies, then multiplied by 100 (30/200 × 100 = 15% recombination)
      (4) Thus, the distance between the rosy eye gene and the curly wing gene is 15 m.u.
   b. To determine the distance between the curly wing gene and the ebony body gene, we need to calculate the percentage of recombination for these two genes
      (1) 160 flies exhibit the parental combination phenotypes (66 + 14 + 12 + 68 = 160)
      (2) 40 flies exhibit the recombinant phenotypes (19 + 2 + 2 + 17 = 40)
      (3) The percentage of recombination is 20% (40/200 × 100)
      (4) The distance between the curly wing gene and the ebony body gene is 20 m.u.
   c. Because map units are additive, the distance between the rosy eye gene and the ebony body gene is 35 m.u. (15 m.u. + 20 m.u.)

**D. Interference**

1. The frequency of double crossovers is expected to equal the product of the frequencies of single crossovers; in the three-factor testcross of Drosophila, it is expected to be 0.20 × 0.15, which equals 0.03
2. However, the observed frequency of double crossovers was 4 out of 200, or 0.02
3. The difference in the observed and expected frequencies of double crossovers is the result of ***interference***
   a. Interference is a measure of the independence of crossover events from one another; the occurrence of one crossover may influence the likelihood of a second crossover
   b. Interference can be positive or negative
4. Interference can be calculated by first determining the ***coefficient of coincidence***

   a. The coefficient of coincidence equals the observed double-crossover frequency divided by the expected double-crossover frequency
   b. Using the three-factor testcross example from above, the coefficient of coincidence equals 0.67
5. Interference is defined as 1 minus the coefficient of coincidence
   a. Using the example above, interference equals 0.33
   b. Therefore, an interference of 0.33 indicates that the actual number of double crossovers will be one-third less than the expected number of double crossovers

### E. Mapping functions

1. The likelihood that double crossovers will occur between linked genes is higher for genes that are more widely spaced than for those that are near one another
2. On average, double crossovers have a recombination frequency of 50%; this is the same recombination frequency that results from single crossovers
3. Accordingly, when map units are determined for gene pairs that are widely spaced, the occurrence of double crossovers will not be detected and the recombination frequency will be underestimated
4. Therefore, when calculated recombination frequencies are greater that 20%, geneticists use a ***mapping function*** to predict map units more accurately

## IV. Advanced Gene Mapping

### A. General information

1. In some organisms, gene mapping is enhanced by fortuitous developmental patterns
   a. In certain fungi and algae, the products of meiosis (spores) remain together in a saclike structure called an *ascus*
   b. Because each product of meiosis contains one chromatid from each tetrad, examination of segregation patterns within asci is called ***tetrad analysis***
2. Gene mapping must be modified in some organisms because of the constraints of the mating systems
   a. For example, gene mapping in humans cannot be based upon three-factor testcrosses
   b. Instead, human gene mapping depends on somatic cell hybridization techniques that rely on the ability to fuse cells in vitro

### B. Gene mapping in haploid fungi

1. In the haploid life cycle, haploid cells fuse to form a meiocyte, which undergoes meiosis to form four haploid products
2. In the *ascomycetes* (fungi that form asci), these products are visible as four spores in one ascus; the spores of some species divide mitotically so that eight spores exist in each ascus (octad)
3. The spores may be arranged in a linear pattern within the ascus (ordered) or they may be randomly distributed (unordered)
4. The distance between two linked genes can be determined by analyzing the tetrads
   a. Consider the hypothetical cross of haploid strains *AB* and *ab*
   b. The diploid meiocyte is *Ab/ab*

c. The resulting asci are categorized into three classes: ***parental ditype (PD), tetratype (TP),*** and ***nonparental ditype (NPD)***
   (1) PD asci contain two types of spores that contain the parental allele combinations (*AB* and *ab*)
   (2) NPD asci contain two types of spores that contain the recombinant allele combinations (*Ab* and *aB* )
   (3) T asci contain all four possible spore types, two of which are parental and two of which are recombinant (*AB, Ab, aB,* and *ab*)
d. If genes *A* and *B* are not linked, PD and NPD asci will occur with equal frequency because of independent assortment
e. If genes *A* and *B* are linked, single crossovers will result in T asci, two-strand double crossovers will result in PD asci, three-strand double crossovers will result in T asci, and four-strand double crossovers will result in NPD asci
f. A recombination frequency can be calculated after gene linkage has been established
   (1) Because PD asci contain all parentals, NPD asci contain all recombinants, and T asci contain one-half recombinants, the recombination frequency equals the number of NPD asci plus one-half of the number of T asci, divided by the number of total asci
   (2) For example, if a testcross yields 74 PD asci, 4 NPD asci, and 22 T asci, the recombination frequency would be $[4 + (22/2)] \div 100$, which equals 0.15

5. In species that give rise to linear tetrads (or octads), geneticists can determine the distance between a gene and the centromere (see *Patterns of Segregation in Fungal Asci,* page 60)
   a. In these species, the two homologs move to opposite ends of a tubelike structure during chromosome separation in meiosis I
   b. When the chromatids separate during meiosis II, the spindles do not overlap; therefore, the two meiotic products at one end of the ascus represent the two chromatids of one homolog, whereas the two meiotic products at the other end of the ascus represent the two chromatids of the other homolog
   c. If no crossing-over occurs in the diploid heterozygote *Aa,* the order of meiotic products in the ascus would be *A-A-a-a* (or *A-A-A-A-a-a-a-a* if the products divide mitotically to form an octad)
      (1) Because the alleles *A* and *a* segregate away from each other during the first division of meiosis, this pattern of segregation is called a first-division segregation pattern or $M_I$ pattern
      (2) Each meiotic product is parental
   d. If crossing-over occurs between the gene and the centromere in the diploid heterozygote *Aa,* the order of meiotic products in the ascus would be one of four types: *A-a-A-a*; *a-A-a-A*; *A-a-a-A*; or *a-A-A-a*
      (1) Because the alleles *A* and *a* segregate away from each other during the second division of meiosis, this pattern of segregation is called a second-division segregation pattern or $M_{II}$ pattern
      (2) Half of the meiotic products in a second-division ascus segregation pattern are parental and half are recombinant
   e. The recombination frequency between the centromere and the *A* locus is therefore equal to half the recombination frequency of the second-division ascus segregation pattern

## Patterns of Segregation in Fungal Asci

$M_I$ and $M_{II}$ patterns of gene segregation are illustrated for two fungal asci. $M_I$ patterns result from a lack of crossing-over between the gene and the centromere (as shown in the top diagram), whereas $M_{II}$ patterns result from crossing-over in this region (as shown in the bottom diagram).

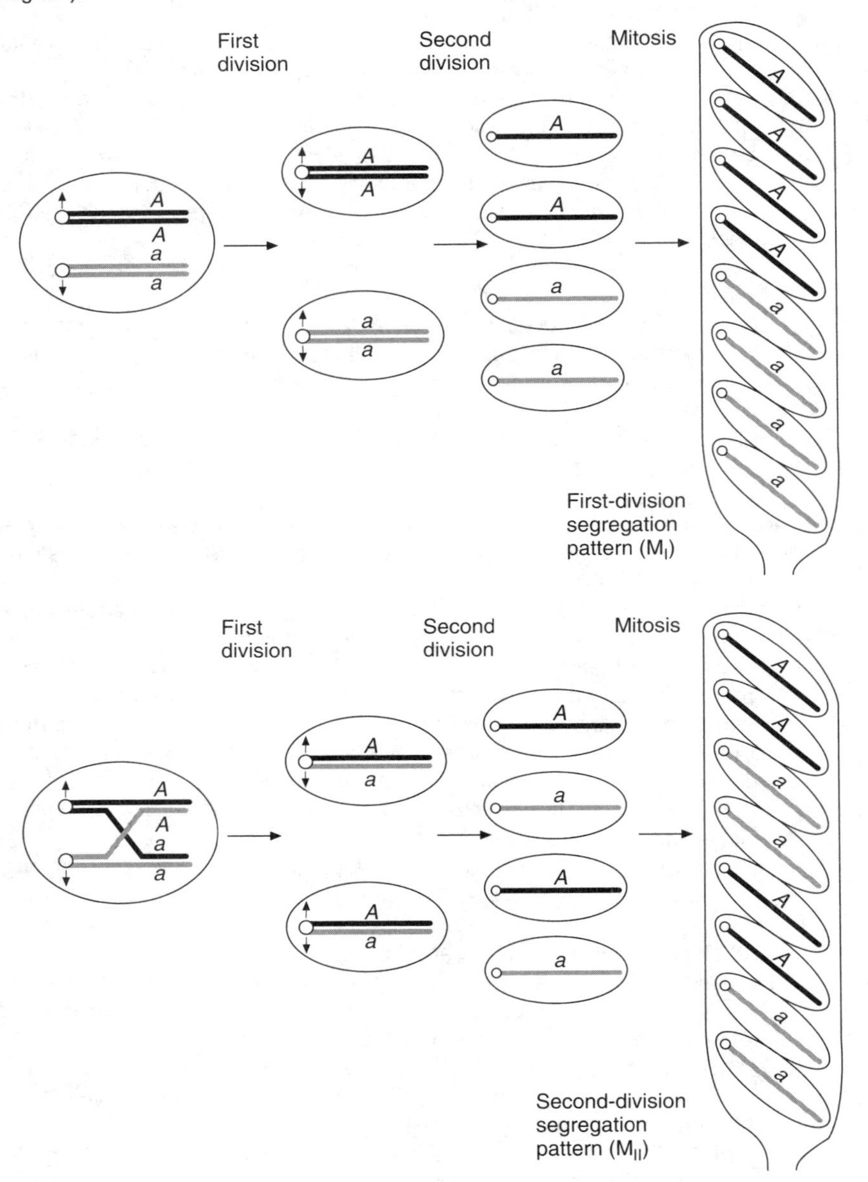

f. For example, if segregation of gene *X* produced 180 first-division asci and 20 second-division asci, the recombination frequency would be (20/200) ÷ 2 which equals 0.05

**C. Gene mapping in humans**

1. Human genes are located on specific chromosomes through the use of human-rodent hybrid cells
   a. Individual cells from human and hamster cell lines (which are cultured in vitro) are fused by growing them together in the presence of polyethylene glycol or specific viral particles
   b. Individual nuclei of fused cells fuse to form a single nucleus that contains human and hamster chromosomes
2. A single-nucleate hybrid cell, or *synkaryon,* will divide mitotically to form a hybrid cell line
   a. For unknown reasons, synkaryons are not completely stable and randomly lose a few human chromosomes during each mitotic division
   b. The loss of human chromosomes continues until only one or a few remain
3. Because of the loss of human chromosomes, different hybrid cell lines will result, each containing different human chromosomes
4. Different cell lines can be screened for the presence or absence of specific human genes, such as those that encode specific enzymes or those that determine specific nutritional requirements
5. Geneticists can draw correlations between the occurrence of certain traits and the occurrence of particular human chromosomes, thus indicating on which chromosome the genes reside
   a. The genes encoding the enzymes lactate dehydrogenase and peptidase (which are located on chromosome 12) were among the first loci to be mapped by this process
   b. The gene encoding coagulation factor III, a blood-coagulating glycoprotein, was assigned to chromosome 1
6. More recent mapping strategies include segregation analysis of DNA fragments (for details, see Chapter 10, The Unit of Function: Molecular Analysis)
7. The most ambitious gene mapping project, called the Human Genome Project, is currently in progress
   a. The objective is to construct a detailed map of the human genome and ultimately determine the entire nucleotide sequence
   b. This project is international in scope and involves geneticists at academic and industrial laboratories
   c. Currently, the human gene map includes approximately 2,000 loci
   d. These include loci associated with breast cancer susceptibility, Duchenne muscular dystrophy, cystic fibrosis, depressive illness, and hemophilia

---

## Study Activities

1. Genes *A* and *B* are located on the same chromosome. Draw a picture of meiosis that illustrates recombination, assuming that the heterozygote undergoing meiosis has the genotype *Ab/aB*.
2. Genes *X* and *Y* are 25 m.u. apart. If an *XY/xy* individual is crossed with an *xy/xy* individual, what progeny genotypes will result and in what proportion?

3. A maize plant of genotype *C Sh Bz/c sh bz* is crossed with a plant of genotype *c sh bz/c sh bz*. The following progeny are produced: 473 *C Sh Bz/c sh bz;* 480 *c sh bz/c sh bz;* 12 *C sh bz/c sh bz;* 15 *c Sh Bz/c sh bz;* 9 *c sh Bz/c sh bz;* 8 *C Sh bz/c sh bz;* 1 *C sh Bz/ c sh bz;* 2 *c Sh bz/c sh bz*. Determine the linear order of loci and the distances that separate them.
4. Genes *L* and *M* are 12 m.u. apart, genes *M* and *N* are 13 m.u. apart, and genes *L* and *N* are 25 m.u. apart. In a three-factor testcross, you observe that the frequency of double crossing-over in this region is 0.011. Calculate the interference.
5. Write a concise paragraph explaining why geneticists may underestimate genetic distances when widely spaced loci are linked.
6. A haploid strain of *Saccharomyces cerevisiae* of genotype *Rs* is mated with a strain of genotype *rS*. After the zygote undergoes meiosis, you observe the following asci: 158 asci contain *Rs* and *rS* spores, 6 asci contain *RS* and *rs* spores, and 36 asci contain *RS, Rs, rS,* and *rs* spores. Are the genes linked? If so, how many map units separate the loci?
7. During an experiment, the human enzyme amylase was found in three human-mouse cell lines: line b, line d, and line e. Line a contains human chromosome 3; line b contains chromosomes 1 and 3; line c contains chromosomes 10 and 20; line d contains chromosomes 1 and 10; and line e contains chromosomes 1 and 20. On which chromosome does the human amylase gene reside?

# 8

# Recombination in Bacteria and Viruses

## Objectives

After studying this chapter, the reader should be able to:

- Differentiate between the processes of conjugation, transformation, and transduction.
- Determine linkage relationships based on interrupted mating experiments.
- Determine gene order based on recombination in merozygotes.
- Describe how transformation experiments and transduction experiments can generate linkage data.
- Illustrate the two life cycles exhibited by bacteriophages.
- Calculate the distance separating two bacteriophage genes using data gathered during mixed infection.

## I. Characteristics of Bacterial Recombination

### A. General information

1. Bacteria contain a single, circular chromosome
2. Bacteria also may contain additional genetic material in the form of a plasmid or episome
    a. A ***plasmid*** is an independent (extrachromosomal) genetic particle that replicates independently of the chromosome
    b. An ***episome*** is a genetic particle that may exist in an independent state, like a plasmid, or may be integrated into the chromosome

### B. Gene transfer

1. Genetic material can be transferred from one bacterial cell to another in one of three ways: conjugation, transformation, and transduction
    a. ***Conjugation*** requires direct cell-to-cell contact, whereas ***transformation*** does not
    b. ***Transduction*** is mediated by bacterial viruses
2. Each of these processes results in the recombination of bacterial genes

## II. Conjugation

### A. General information

1. In 1946, Joshua Lederberg and Edward Tatum discovered that strains of bacteria could transfer genetic material from one to another

a. Different ***auxotrophs*** — strains unable to grow in the absence of a nutritional supplement — were grown together
b. ***Prototrophs*** — strains able to grow in the absence of a nutritional supplement — appeared at a low frequency (1 in 10 million)
c. Lederberg and Tatum concluded that the auxotrophs produced a recombinant strain by exchanging genetic material
2. Cell-to-cell contact was required for this type of genetic exchange

**B. The fertility factor**
1. Conjugation is a one-way exchange of genetic material
a. DNA is transferred from a donor cell to a recipient cell
b. Physical contact between donor and recipient cells is mediated by hairlike structures called *pili*
2. A cell's ability to form pili and exchange genetic material is conferred by the ***fertility factor (F)***
a. F is a small, circular episome that contains approximately 100 genes
b. F can replicate independently or become integrated into the bacterial chromosome
c. $F^+$ cells are donors; $F^-$ cells are recipients

**C. Gene transfer**
1. If F exists in a nonintegrated, autonomous state, it is transferred in its entirety from the donor to the recipient
a. When conjugation is initiated, the double-stranded F molecule begins replication
(1) A DNA nick (a break in only one strand) occurs at the origin
(2) The nicked strand is transferred into the donor cell
(3) F is transferred in a linear fashion, beginning with the origin and ending with the terminus
(4) As one strand of the DNA double helix is transferred, the donor cell synthesizes a new strand to replace it
(5) When the transfer is completed, the recipient cell synthesizes a complementary copy of F, thereby creating a double-stranded molecule
b. Both exconjugants (that is, the donor and recipient cells) contain F
2. If F is integrated into the bacterial chromosome at the time of conjugation, F is not transferred in its entirety; instead, it drags along chromosomal genes (see *Bacterial Conjugation*)
a. A bacterial strain in which F is present in an integrated form is called an ***Hfr*** strain; the "Hfr" refers to high frequency of recombination in this strain
b. An Hfr strain is formed when a single crossing-over occurs between the bacterial chromosome and F, resulting in a single, circular, recombinant molecule
c. F no longer is replicated and transferred independently; during conjugation, genetic transfer is initiated at the F origin and continues around the entire bacterial chromosome in the direction of the F terminus
d. Because of random cell movement, conjugation typically is interrupted before the transfer of the entire recombinant molecule is complete

## Bacterial Conjugation

The diagrams below show how genetic information is transferred during conjugation. The diagram on the left shows the transfer of an autonomous fertility factor (F) in an $F^+$ and $F^-$ mating. The diagram on the right shows the transfer of an integrated fertility factor and chromosomal genes in an *Hfr* and $F^-$ mating.

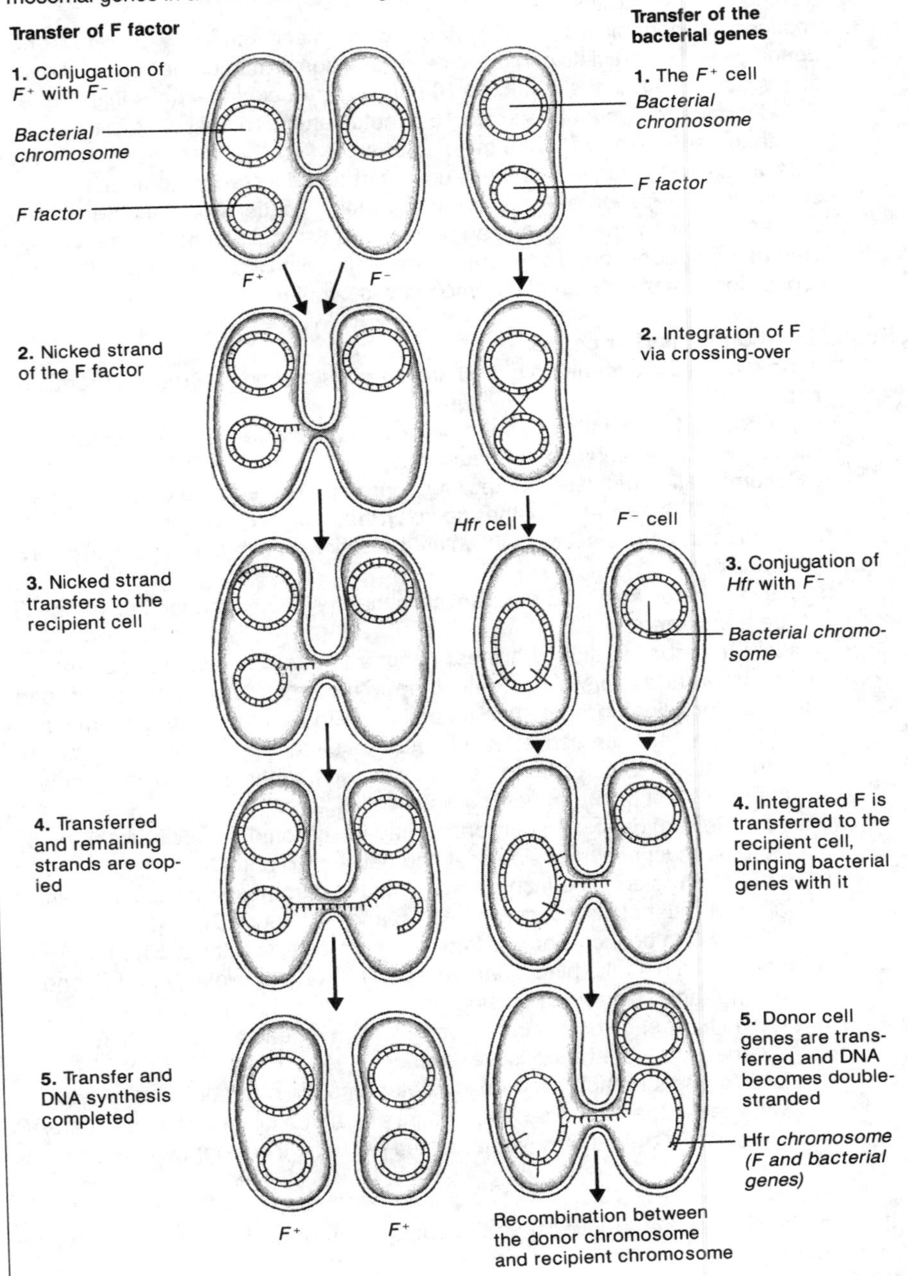

**D. Interrupted mating experiments**

1. Because gene transfer occurs in a linear manner, the chromosomal genes that enter the recipient cell first are the genes located adjacent to the site of F integration
2. The time between the transfer of chromosomal genes is proportional to the distance that separates them
3. If matings are periodically interrupted by agitation and screened for transfer of donor genes, geneticists can derive a chromosome map of the bacterial cell
   a. If donor gene *A* is first detected 10 minutes after conjugation is initiated, and donor gene *B* is first detected 15 minutes after conjugation is initiated, then genes *A* and *B* are 5 minutes apart
   b. Thus, minutes are the map units used for bacterial chromosome maps
4. Because the integration of F can occur at various locations on the chromosome, different chromosome regions can be mapped using different Hfr strains
5. Comparison of gene maps constructed from different Hfr strains led to the conclusion that the bacterial chromosome is circular

**E. Recombination of marker genes**

1. Interrupted mating experiments facilitate the construction of a rough chromosome map
2. A more detailed map that shows the order of closely linked genes is created by calculating the recombination frequencies
   a. Recombination between the recipient chromosome and the donor fragment occurs after formation of the ***merozygote,*** a partially diploid cell
   b. It is important to select cells in which the entire region of interest has been transferred
   c. This is done by selecting recipient cells that have acquired the last gene to be transferred
3. Consider the recombination of a linear donor fragment with alleles *A, B,* and *C* (in that order) and a circular chromosome with corresponding alleles *a, b,* and *c*
   a. In order for a linear fragment to recombine with a circular chromosome, an even number of crossover events must take place
   b. For the entire *A-B-C* region to be incorporated into the chromosome, crossing-over must occur on both sides of the region
   c. For the *A-B* region to be incorporated into the chromosome without *C,* crossing-over must occur outside of *A* and between *B* and *C*
   d. For the *B-C* region to be incorporated into the chromosome without *A,* crossing-over must occur between *A* and *B* and outside of *C*
   e. For *A* and *C* to be incorporated into the chromosome without *B,* four crossing-overs must take place: one outside of *A,* one between *A* and *B,* one between *B* and *C,* and one outside of *C*
      (1) The frequency of four crossover events will be much lower than the frequency of two crossover events
      (2) This phenomenon allows us to determine the order of linked genes
      (3) Because the genotype *AbC* occurs at a much lower frequency than *ABc* and *aBC,* gene *B* must be located between the other two genes

**F. Sexduction**

1. F sequences can be removed from the bacterial chromosome by looping-out and crossing-over

2. Occasionally, the excision event is not perfect and results in an F molecule that contains a small region of the host chromosome
3. Transfer of the aberrant F molecule, called F′, results in the high-efficiency transfer of a specific chromosomal region
4. This specialized form of conjugation is called ***sexduction*** or F′-duction
5. Sexduction therefore involves the infectious transfer of bacterial genes
   a. Infectious transfer of genes that confer antibiotic susceptibility leads to the rapid evolution of antibiotic-resistant strains of bacteria
   b. This phenomenon is particularly threatening in hospitals and other health care facilities

## III. Transformation

### A. General information

1. Transformation was discovered by Frederick Griffith in 1928, and ultimately led to the conclusion that deoxyribonucleic acid (DNA) is the genetic material
2. Transformation occurs through the uptake of exogenous DNA (that is, DNA from an external source)
3. Physical contact of donor and recipient cells is not required for transformation to occur

### B. Mechanisms of transformation

1. Exogenous, double-stranded DNA binds to receptor sites on the cell surface during certain stages of the cell cycle
2. Once bound, DNA can be taken into the cell
3. The DNA fragment is integrated into the host chromosome

### C. Linkage data obtained from transformation

1. Genes that are closely linked have a greater chance of being carried on the same fragment of transforming DNA
2. Closely linked genes will therefore have a higher frequency of double transformation
3. The frequency of double transformation of any two genes serves as an estimate of the distance between them
4. The mapping of loci based upon the frequencies of recombination among three linked genes also can be performed in transformation experiments (as in conjugation experiments)

## IV. Transduction

### A. General information

1. Bacteriophages (or phages) are viruses that infect bacteria and carry small fragments of bacterial DNA from one host to another
2. This process, which facilitates bacterial recombination, is called transduction

### B. Bacteriophage life cycles

1. Phage particles have a protein coat that surrounds the genetic material
2. Phages exhibit two basic life cycles: ***lytic*** and ***lysogenic***

a. The lytic cycle involves attachment of phage particles to the bacterium, injection of viral genetic material into the host, breakdown of the bacterial chromosome, replication and expression of phage genetic material, assembly of phage progeny, and release of the progeny by lysis of the host cell
b. The lysogenic cycle includes the integration of phage genetic material into the bacterial chromosome prior to the replication and expression of phage genetic material
(1) An integrated phage is called a ***prophage***
(2) The prophage will be replicated along with the bacterial chromosome as long as it remains integrated
(3) A prophage can be induced to excise, at which time it enters the lytic cycle
(a) Various environmental conditions, such as ultraviolet light, can induce the lytic cycle
(b) The excision process involves the outlooping of prophage DNA followed by a crossover event between the ends of the prophage DNA, thus releasing the integrated sequence

3. Some phages exhibit only the lytic cycle (such as phage T2, phage T4, and other virulent ones); others exhibit either life cycle (such as phage lambda and other temperate phages)

**C. Generalized transduction**

1. During the production of phage progeny in the lytic cycle, a small fragment of partially degraded bacterial chromosome may be accidentally incorporated into the protein coat of a phage particle
2. This particle, called a transducing phage, can inject bacterial DNA into another cell, called a transductant
3. This entire process, which is called generalized transduction, facilitates the transfer of any bacterial gene, albeit at a low frequency
4. Closely linked genes may be carried on the same fragment of bacterial DNA, leading to cotransduction
5. The frequency of cotransduction of any two genes serves as an estimate of the distance between them
6. The mapping of loci based upon the frequencies of recombination among the three linked genes also can be performed in transduction experiments

**D. Specialized transduction**

1. Some phages that are capable of lysogeny (that is, entering the lysogenic cycle) can integrate into the host chromosome only at specific locations
a. Phage lambda ($\lambda$), for example, integrates into the bacterial chromosome at the *att* $\lambda$ site
b. A single crossing-over between the bacterial *att* $\lambda$ site and the phage *att* site results in the incorporation of the circular phage lambda chromosome
2. Occasionally, the prophage excises abnormally when it enters the lytic cycle, producing both a phage particle that contains a fragment of the bacterial chromosome adjacent to the incorporation site and a bacterial chromosome that contains a piece of the phage DNA
a. Phage lambda, for example, occasionally incorporates the adjacent bacterial genes *gal* (galactose fermentation) or *bio* (biotin biosynthesis) to form $\lambda$*dgal* or $\lambda$*dbio,* respectively (*d* indicates a defective phage)

b. Although specialized transducing phage particles are defective and cannot integrate independently into a new host chromosome, they can do so in the presence of a helper phage that provides the required gene products

# V. Bacteriophage Crosses

## A. General information

1. In addition to the use of bacteriophages as tools for the mapping of bacterial genes, linkage analyses of bacteriophage genes also can be performed
2. A mixed infection of host cells with two phage genotypes results in the production of recombinant phage

## B. Plaque phenotypes

1. Lysis of host cells during the phage lytic cycle can be visualized as ***plaques,*** which appear as clear areas on the bacterial lawn of a petri dish
2. The morphology of a plaque (size, clarity, and so on) depends on the genotype of the infecting phage
3. The ability to initiate the phage life cycle also depends on the infecting phage's genotype
   a. The host range of a phage is defined as those bacterial strains that are not immune to infection
   b. The presence or absence of plaques after cell infection can be used to determine the host range genotype

## C. Linkage data obtained from mixed infections

1. In 1949, Alfred Hershey performed crosses using two phage T2 genotypes: $h^-r^+$ and $h^+r^-$
   a. An $h^-$ phage can infect two strains of bacteria, thereby giving rise to clear plaques (which indicate complete lysis) when infection occurs on petri dishes containing a bacterial lawn that has both strains
   b. An $h^+$ phage can infect only one strain of bacteria, giving rise to cloudy plaques because of the phage's ability to lyse only one strain of cells when infection occurs on petri dishes containing a bacterial lawn that has both strains
   c. An $r^+$ phage can produce small plaques as a result of the slow lysis of infected cells
   d. An $r^-$ phage can produce large plaques as a result of the rapid lysis of infected cells
2. After a double infection was performed, the cell lysate that contained the progeny phage was photographed and individual plaque phenotypes were determined
   a. Four genotypes were evident: $h^-r^+$ (clear, small plaques), $h^+r^-$ (cloudy, large plaques), $h^-r^-$ (clear, large plaques), and $h^+r^+$ (cloudy, small plaques)
   b. The frequency of recombination was determined to be the number of recombinant plaques ($h^-r^-$ and $h^+r^+$) divided by the total number of plaques

3. By performing linkage analyses with different gene pairs, the phage T2 linkage map was found to be circular

## Study Activities

1. Make a table that summarizes the major characteristics and methods of linkage analysis associated with each of the three types of bacterial genetic transfer
2. A transformation experiment is performed using donor genotype *XYZ* and recipient genotype *xyz*. The following 1,000 colony genotypes result: 566 *xyz*, 93 *Xyz*, 100 *xYz*, 104 *xyZ*, 8 *XYz*, 42 *XyZ*, and 87 *xYZ*. What is the probable gene order? Which two genes are most closely linked?
3. You perform *Escherichia coli* interrupted mating experiments with each of three Hfr strains. Strain 1 gives a gene order of *A-E-C-D*. Strain 2 gives a gene order of *C-D-F-B*. Strain 3 gives a gene order of *F-B-A-E*. What is the order of these genes on the circular *E. coli* chromosome?
4. Determine the order for genes *L, M, N,* and *O* based on the following percentages of cotransduction: *L-M*, 0%; *M-N*, 33%; *M-O*, 27%; *L-O*, 3%; *L-N*, 1%.
5. A phage cross of *abc* and $a^+b^+c^+$ yields the following progeny: 348 *abc*, 352 $a^+b^+c^+$, 81 $abc^+$, 17 $ab^+c$, 48 $ab^+c^+$, 42 $a^+bc$, 13 $a^+bc^+$, and 99 $a^+b^+c$. Calculate gene distances and draw a gene map.

# 9

# The Unit of Function: Classical Analyses

## Objectives

After studying this chapter, the reader should be able to:

- Diagram biochemical pathways based on the response of mutants to nutritional supplementation.
- Explain the relationship between genes and enzymes.
- Predict the genetic behavior of mutant alleles based on the resulting physiological defects.
- Utilize *cis-trans* tests for determining units of function.
- Draw detailed maps based on intragenic recombination.

## I. One Gene – One Enzyme Hypothesis

### A. General information

1. Although Mendel recognized that an organism's phenotype was the product of its genotype, the method by which genes exert their effects was not known until 1941, when G. Beadle and E. Tatum concluded that genes control the production of enzymes
    a. An *enzyme* is a catalyst that increases the rate of biochemical reactions
    b. Beadle and Tatum recognized that enzymes regulate the conversion of one metabolic intermediate into another
        (1) For example, the sugar glucose is broken down into carbon dioxide and water through a series of reactions that collectively is called cellular respiration
        (2) Each step of cellular respiration depends upon the action of a specific enzyme
    c. They hypothesized that each enzyme was controlled by a single gene
2. The one gene – one enzyme hypothesis united the fields of genetics and biochemistry

### B. Overview of biosynthetic pathways

1. A biochemical pathway leading to the production of a specific product can be viewed as a series of molecular conversions, each of which is catalyzed by a single enzyme
2. Consider a hypothetical product (P), which is formed from compound 1; compound 2, in turn, is formed from compound 1
    a. These reactions may be symbolized as: $1 \rightarrow 2 \rightarrow P$

b. Let us assume that the 1 → 2 reaction is catalyzed by enzyme A and that the 2 → P reaction is catalyzed by enzyme B
c. Enzymes A and B are specified by genes *a* and *b,* respectively
3. If gene *a* or *b* is inactivated, the corresponding enzyme will not be present and the biosynthetic pathway is blocked
   a. If gene *a* is inactivated, compound 2 and P will not be produced, while compound 1 will accumulate
   b. If gene *b* is inactivated, P will not be produced, while compound 2 will accumulate
4. If a cell in which a gene has been inactivated is provided with an exogenous source of the metabolic intermediate that normally is produced, then its product ultimately may be formed and the cell may survive
   a. If a cell in which gene *a* is inactive is provided with an exogenous source of compound 2, then the metabolic block is bypassed and the cell will produce P
   b. However, if a cell in which gene *a* is inactive is provided with an exogenous source of compound 1, the metabolic block is not bypassed and the cell will not produce P
5. In summary, added metabolic compounds will support the growth of mutants that are blocked in steps prior to the formation of the added compound; the compound that supports the growth of the most mutants is produced last in the pathway
6. Beadle and Tatum used this logic to study the arginine biosynthetic pathway of fungi of the genus *Neurospora*

**C. Experiments of Beadle and Tatum**

1. Three different auxotrophic strains of *Neurospora* were examined; each strain required arginine supplementation for growth
   a. Each strain appeared to contain a single mutant gene
   b. Each mutant gene was mapped to a different locus (designated here as *arg-1, arg-2,* and *arg-3*)
2. Each mutant strain was studied for its ability to grow in the presence of ornithine and citrulline, two precursors of arginine
   a. The *arg-1* mutant gene grew in the presence of ornithine or citrulline
   b. The *arg-2* mutant gene grew in the presence of citrulline, but not ornithine
   c. The *arg-3* mutant gene did not grow in the presence of ornithine or citrulline
3. Beadle and Tatum concluded that the *arg* mutant genes represented three separate genes; these, in turn, give rise to three different enzymes that catalyze arginine production
   a. The wild-type *arg-1* allele catalyzes the production of ornithine from a precursor molecule
   b. The wild-type *arg-2* allele catalyzes the production of citrulline from ornithine
   c. The wild-type *arg-3* allele catalyzes the production of arginine from citrulline
4. The one gene – one enzyme hypothesis therefore implies that biochemical reactions occur in discrete steps, each of which is catalyzed by a single enzyme that is controlled by a single gene (see *Arginine Biosynthesis*)

**D. The gene-protein relationship**

1. The connection between genes and enzymes was elucidated by Vernon Ingram in 1957

## Arginine Biosynthesis

The top diagram illustrates the arginine biosynthetic pathway. The bottom chart shows the response of arginine auxotrophs to supplements.

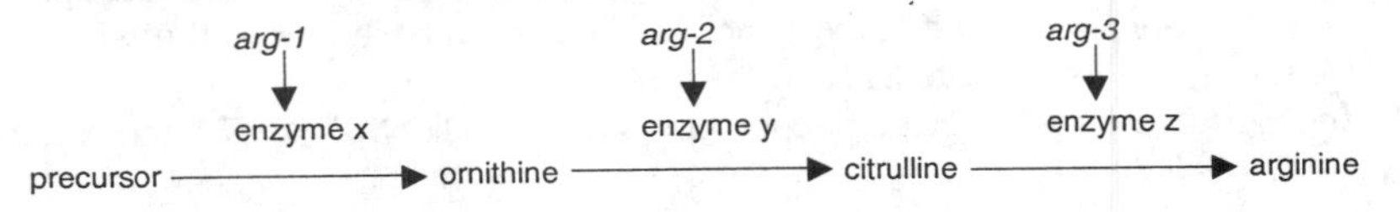


| Nutritional supplement | mutation | | |
|---|---|---|---|
| | *arg-1* | *arg-2* | *arg-3* |
| ornithine | growth | no growth | no growth |
| citrulline | growth | growth | no growth |
| arginine | growth | growth | growth |

From *An Introduction to Genetic Analysis* (5th ed.) by Suzuki, Griffiths, Miller, Lewontin, and Gelbart. ©1993 by W.H. Freeman and Company. Reprinted with permission.

a. Most enzymes are proteins; however, some RNA molecules act as enzymes

b. Proteins are macromolecules consisting of linear arrays of amino acids called polypeptides

(1) The linear order of amino acids is called the primary structure of a protein

(2) Localized folding of the amino acid chain, which is stabilized by hydrogen bonding, is called the secondary structure of a protein

(3) The overall, three-dimensional shape of an amino acid chain is called the tertiary structure of a protein

(4) The binding of two or more amino acid chains to form a polymer is called the quaternary structure of a protein

2. Ingram isolated hemoglobin molecules from healthy individuals and from those with sickle cell anemia

   a. Molecules from both types of individuals were subjected to *protein sequencing,* the determination of amino acid order within the protein

   b. Individuals with sickle cell anemia had hemoglobin molecules that contained a single amino acid alteration

3. He concluded that genes specify the primary structure of a single polypeptide

4. Geneticists now know that genes specify proteins via an RNA intermediate called ***messenger RNA (mRNA);*** gene sequences specify mRNA sequences, which, in turn, specify protein sequences

## II. Biochemical Explanations for Mendelian Genetics

### A. General information

1. After having examined the nature of the gene and its role in biochemical pathways, we can reexamine Mendelian principles and seek explanations that are consistent with cellular metabolism
2. Knowledge of physiological mechanisms helps predict the genetic behavior of mutant alleles

### B. Dominance and recessiveness

1. Dominance generally results from enzyme function
2. Recessiveness generally results from lack of enzyme function
   a. For example, individuals with the recessive disorder phenylketonuria lack the enzyme phenylalanine 4-hydroxylase, which converts the amino acid phenylalanine into the amino acid tyrosine
   b. Individuals with the recessive disorder galactosemia lack galactose-1-phosphate uridyltransferase, an enzyme needed to metabolize the sugar galactose
3. For a diploid cell or organism to completely lack enzyme activity, both copies of the gene that control the enzyme must be mutant; hence, a recessive mutation must be homozygous if it is expressed
4. In heterozygosity, one dominant allele that confers normal enzyme activity usually is sufficient for full cellular functioning
5. If a metabolic defect is caused by a dominant allele, the enzyme function of that cell generally has been altered — not eliminated

### C. Mendelian ratios

1. Chapter 3, Advanced Mendelian Analysis, defines several types of gene interaction that produce modified $F_2$ dihybrid ratios, including epistatic, complementary, and duplicate gene action
2. Specific biochemical circumstances can explain each of these types of gene interaction
   a. An epistatic gene may alter a biochemical pathway prior to another alteration caused by a second gene
   b. Complementary gene action can be explained by a single, two-step biochemical pathway (or two pathways), in which dominant alleles (that is, the functional enzymes) are required at both loci for the making of the product and the expression of the associated phenotype
   c. Duplicate gene action can be viewed as a duplication of enzyme function; the functional enzyme can be specified by either of two loci

## III. Complementation

### A. General information

1. A variant phenotype caused by a single-gene mutation is mediated by a change in a single polypeptide
2. Because a specific phenotype generally is determined by the interaction of several polypeptides, one logically can ask if it is possible to determine whether two

variant phenotypes represent different changes in a single gene (different alleles) or changes in two distinct genes

3. If a wild-type phenotype results from the cross of two mutants, the mutations are said to complement one another; this is called ***complementation***
   a. Complementation implies that the two mutations occur in two different genes
   b. A lack of complementation implies that the two mutations occur in the same gene and therefore represent different alleles
4. Genes are defined as units of function based on ***complementation (cis-trans) tests***

**B. The complementation test**

1. Complementation, or *cis-trans*, tests are conducted by combining two mutations in the same cell or organism (see *Cis-trans Test,* page 76 )
   a. The term "*cis*" literally means "on this side"; mutations in the cis configuration are located on the same chromosome
   b. The term "*trans*" refers to "the other side"; mutations in the trans configuration are located on different homologs
2. The mutations are combined in the *trans* and *cis* configurations
   a. The *trans* configuration is produced by bringing together two mutations from different parents
      (1) In diploid cells or organisms, this can be accomplished by crossing together two individuals that are homozygous for a different mutation
      (2) In prokaryotes or bacteriophages, a diploid (or partial diploid) condition can be achieved using techniques described in Chapter 8, Recombination in Bacteria and Viruses, so that the mutations are introduced on different chromosomes
      (3) If the *trans* configuration results in the production of a wild-type phenotype, the mutations must be on different genes
      (4) If the *trans* configuration results in the production of a mutant phenotype, then the mutations are on the same gene
   b. The *cis* configuration is produced by introducing two mutations from the same parent into a common cell or organism
      (1) The *cis* configuration serves as a control and always should yield the wild-type phenotype
      (2) Dominant mutations, therefore, cannot be used in *cis-trans* tests
3. Consider, for example, the marriage of two albino individuals
   a. Albinism may be the result of mutation at several different loci; two recessive *a* genes are required for albinism
   b. Children of the marriage will have their parents' mutations in the *trans* configuration of the appropriate gene and may or may not exhibit complementation
      (1) If the two albino parents give birth to an albino child (without complementation), then the parents' albinism results from mutations on the same gene (*aa* and *aa* mating produces *aa*)
      (2) If the two albino parents give birth to a normal child (indicating complementation), then the parents' albinism results from mutations in different genes ($A_1A_1a_2a_2$ and $a_1a_1A_2A_2$ mating produces $A_1a_1$ $A_2a_2$)
4. The *cis-trans* test occasionally gives misleading results
   a. The *trans* configuration sometimes exhibits complementation when both mutations are in the same gene; this is called *intragenic complementation*

**Cis-trans Test**

Mutation sites are indicated by dark squares in genes *A* and *B*. A + indicates complementation; a – indicates no complementation.

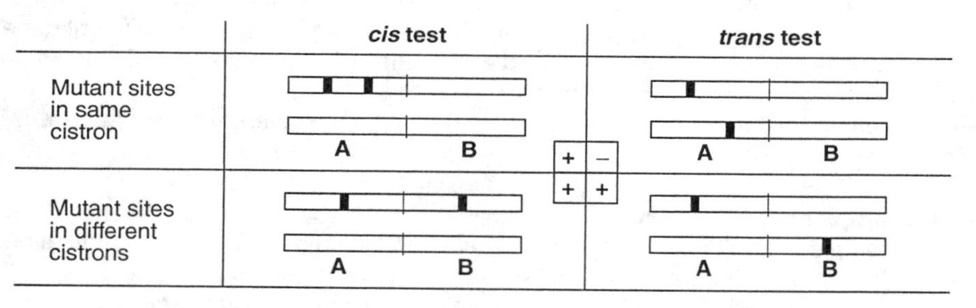


From *An Introduction to Genetic Analysis* (5th ed.) by Suzuki, Griffiths, Miller, Lewontin, and Gelbart. ©1993 by W.H. Freeman and Company. Reprinted with permission.

b. Intragenic complementation arises when two defective polypeptides, each arising from a different mutant allele, interact in such a way that full or partial activity is restored

5. The unit of function defined by the *cis-trans* test is called the ***cistron***
   a. The cistron is the functional definition of the gene
   b. The term cistron is used interchangeably with the term gene, although the former term is falling into disuse

## IV. Intragenic Recombination

### A. General information

1. Geneticists wondered if the cistron, the smallest unit of function, also is the smallest unit of recombination
   a. Prior to the work of Seymour Benzer in 1959, the prevailing view of recombination was that chromosome breakage occurred between genes
   b. The gene itself (the cistron) was considered indivisible
2. By screening very large numbers of phage progeny in which two mutations in the same gene were in the *trans* configuration, Benzer detected a low frequency of ***intragenic recombination***
   a. Intragenic recombination between two mutation sites in the same gene gives rise to one allele that carries both mutations and one allele that is normal
   b. Intragenic recombination is possible because the gene is comprised of many nucleotide pairs that are divisible by recombination

### B. Ordering of mutational sites

1. The mapping of mutation sites within a gene can be performed using the calculation of map distances

a. The number of recombinant progeny divided by the total number of progeny is equal to the recombination frequency
b. Because one of the recombinant products (the double mutant) resembles the parental products (single mutants), the total number of recombinants is calculated as twice the number of wild-type recombinants

2. Mutation sites also can be ordered by examination of the flanking genes
   a. If an individual with one mutation in gene *X* ($x_1$) that is flanked by genes *A* and *B* is crossed with an individual with a second mutation in gene *X* ($x_2$) that is flanked by genes *a* and *b*, examination of wild-type recombinants will reveal gene order
   b. If the wild-type recombinant contains *a* and *B*, then the site of mutation $x_1$ is to the left of the site of mutation $x_2$ (that is, closer to gene *A*)
   c. If the wild-type recombinant contains *A* and *b*, then the site of mutation $x_1$ is to the right of the site of mutation $x_2$ (that is, closer to gene *B*)

### C. Deletion mapping

1. The mutation sites discussed thus far represent ***point mutations,*** a mutation in a single nucleotide pair
2. A ***deletion*** is a type of mutation that results in the loss of several or many nucleotide pairs
3. Sometimes intragenic recombination can occur between deletions as well as between point mutations and deletions
   a. If intragenic recombination occurs between two deletions, the regions encompassed by the two deletions do not overlap
   b. If intragenic recombination does not occur between two deletions, the regions encompassed by the two deletions overlap
   c. If intragenic recombination occurs between a point mutation and a deletion, the point mutation lies outside of the region encompassed by the deletion
   d. If intragenic recombination does not occur between a point mutation and a deletion, the point mutation lies within the region encompassed by the deletion
4. Deletion mapping can accelerate gene mapping by rapidly locating genes on a chromosome segment
5. In humans, deletions are associated with cri-du-chat syndrome, Wolf-Hirschhorn syndrome, Wilms' tumor, retinoblastoma, and Prader-Willi syndrome

## Study Activities

1. You have identified mutations in three genes (mutant 1, 2, and 3). Each mutant requires compound W for growth. Compounds X, Y, and Z occur in the same biochemical pathway as compound W. Compound X supports growth of mutant 2. Compound Y does not support growth of any mutant. Compound Z supports growth of mutants 1 and 2. Determine the order of compounds in the pathway and the points at which each mutant is blocked.

2. Phenylketonuria is a metabolic disorder resulting from a deficiency in the enzyme phenylalanine hydroxylase. Is this disorder most likely caused by a recessive or dominant allele? Explain your answer.
3. Eye color in *Drosophila* is the result of two pigments: a brown pigment and a vermilion pigment. Brown mutants *(bw)* do not produce the vermilion pigment because of loss of enzyme activity. Vermilion mutants *(v)* do not produce the brown pigment because of loss of enzyme activity. A homozygous vermilion female is crossed with a homozygous brown male (vermilion is X-linked; brown is autosomal). What are the genotypes and phenotypes of the resulting $F_1$ and $F_2$ progeny?
4. Alkaptonuria and tyrosinosis are recessive metabolic disorders caused by the inability to metabolize phenylalanine. An individual who has alkaptonuria and an individual who has tyrosinosis have a normal child. Based on this information, what type of mutations cause alkaptonuria and tyrosinosis in these individuals?
5. Four different deletion mutants are intercrossed in all possible combinations to test for intragenic recombination. Intragenic recombination is not observed between deletions 1 and 2, between deletions 1 and 4, or between deletions 3 and 4. Draw the chromosomal region in question and indicate the linear order and relative positions of the four deletions.

# 10

## The Unit of Function: Molecular Analyses

### Objectives

After studying this chapter, the reader should be able to:

- Explain the concept of restriction mapping.
- List and describe the enzymes used in recombinant deoxyribonucleic acid (DNA) research.
- Explain the procedure for genomic cloning.
- Compare and contrast different cloning vectors.
- Describe the various levels of selection required to identify a specific gene clone after constructing a genomic library.
- Explain the uses of complementary DNA cloning.
- Compare and contrast the procedures and usefulness of Southern and Northern blotting.
- Explain the concepts of restriction fragment length polymorphisms and polymerase chain reaction, and how they are useful in gene analysis.
- Draw a flow diagram that illustrates DNA sequencing.

### I. Introduction to Gene Cloning

**A. General information**

1. For high-resolution study of gene structure, geneticists rely upon the technique of gene cloning
    a. The term cloning refers to the production of identical copies
    b. Gene cloning, or molecular cloning, refers to the production of many identical copies of a specific DNA sequence for analysis
2. Gene cloning requires the production of recombinant DNA, the cutting and joining of DNA molecules such that the sequences to be cloned are incorporated into a self-replicating molecule

**B. Basic techniques**

1. The self-replicating molecule into which passenger DNA is inserted, such as a bacterial plasmid, is called a *vector*
2. The vector must be purified and enzymatically cut open before it can accept passenger DNA
3. The vector and the passenger DNA are enzymatically joined to form a recombinant DNA product

4. Recombinant DNA is introduced into a host cell, typically *Escherichia coli,* for propagation of the passenger DNA

## II. Restriction Endonucleases

### A. General information

1. Cutting DNA at specific sites for the creation of recombinant molecules requires the use of specialized enzymes called ***restriction endonucleases,*** or restriction enzymes
2. An endonuclease is an enzyme that catalyzes the breakage of the phosphodiester linkages in DNA
3. Restriction endonucleases create double-stranded breaks in DNA molecules
4. In vivo, these enzymes protect bacterial cells from foreign DNA invasion
5. Restriction enzymes are named after the organisms from which they are isolated; for example, enzyme *Eco*RI was discovered in *E. coli*

### B. Site specificity

1. Restriction endonucleases cut DNA at or near specific sites called ***restriction sites***
2. Restriction sites contain specific nucleotide sequences
   a. They typically are palindromic — that is, the sequence on one DNA strand is the same as the sequence on the other strand, but in reverse
   b. For example, the restriction enzyme *Eco*RI recognizes and cuts DNA at the 5′-GAATTC-3′ sequence; the complementary strand of DNA contains a 3′-CTTAAG-5′ sequence at that site
   c. Restriction sites commonly contain 4, 6, or 8 nucleotide pairs
3. Restriction endonucleases may produce blunt-ended molecules or may produce single-stranded overhangs caused by a staggered cut
4. Restriction endonucleases that produce single-stranded overhangs at palindromic restriction sites also produce complementary ends

### C. Restriction mapping

1. Restriction sites can serve as molecular landmarks in the physical analysis of DNA molecules; the location of cleavage sites is called ***restriction mapping***
2. The locations of restriction sites are determined by cutting DNA with restriction endonucleases and observing the size of the resulting fragment (see *Restriction Mapping*)
   a. DNA is cut with single enzymes and with combinations of two enzymes
   b. The resulting fragments are separated by *electrophoresis,* the migration of charged molecules in a particular medium when passed through an electric field
      (1) The electrophoresis of DNA fragments uses a gel-like matrix of agarose or polyacrylamide
      (2) Small molecules move more rapidly through the gel than large molecules because they are less affected by the density of the matrix; this results in the separation of DNA fragments according to size
      (3) For accurate size determination, geneticists compare the migration of unknown DNA fragments with the migration of DNA fragments of known length

### Restriction Mapping

Electrophoresis of DNA fragments generated by digestion with one or two restriction endonucleases facilitates the determination of restriction sites relative to one another. The top illustration depicts autoradiography of DNA fragments after electrophoresis. The bottom illustration shows the restriction site map.

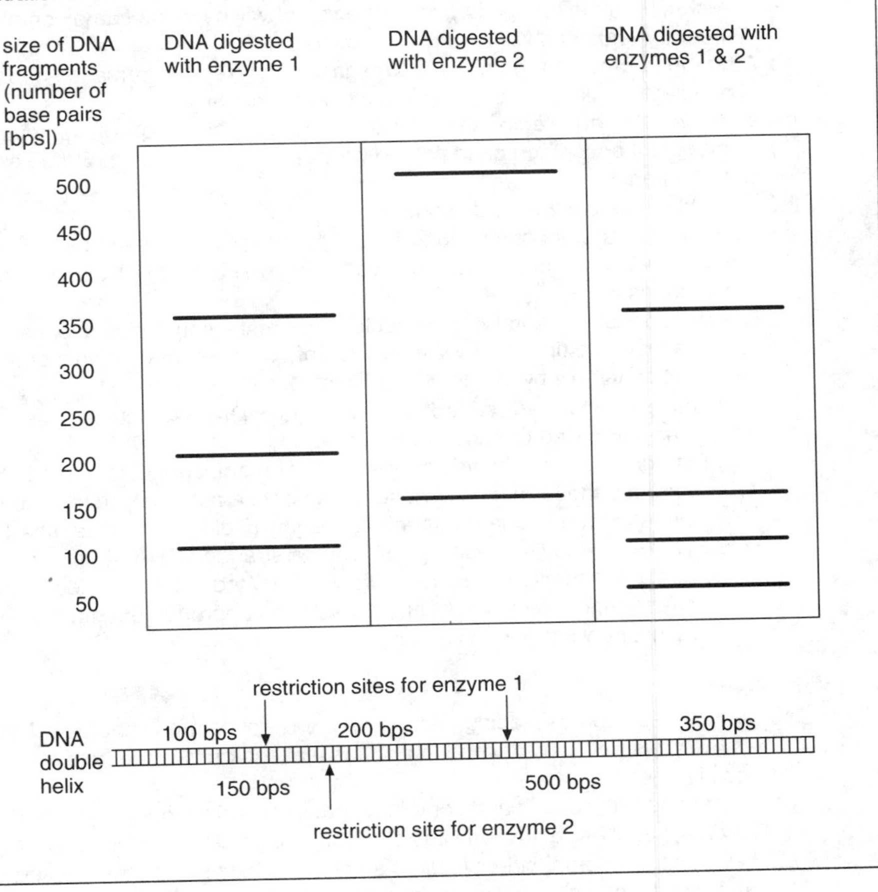


## III. Recombinant DNA

### A. General information

1. After they are generated by restriction endonucleases, DNA fragments can be joined together in various combinations to produce recombinant DNA molecules
2. Recall that for gene cloning, DNA fragments are inserted into the vector that has been cut open by restriction endonucleases
3. Covalent linkage of DNA fragments in vitro requires the enzyme called DNA ligase (for details about DNA ligase, see Chapter 5, Structure and Replication of Genetic Material)

**B. Ligation**

1. The single-stranded ends of DNA fragments that have overhangs (as a result of staggered cutting) are complementary to one another; these complementary ends are called sticky ends because they tend to form base pairs
   a. If the DNA to be cloned is cut with the same enzyme that is used to cut the vector, then the foreign DNA and the vector will have the same complementary ends and will tend to form base pairs
   b. After the fragments are paired, DNA ligase catalyzes the formation of phosphodiester linkages, resulting in a single molecule
2. If the foreign DNA and vector are cut with different enzymes, or with enzymes that generate blunt ends, then base pairing will not occur and ligation will be extremely limited
3. Incompatible ends can be made compatible
   a. If sticky ends are not compatible, the single-stranded overhangs are digested with an exonuclease specific to single-stranded DNA, resulting in blunt ends
   b. The blunt ends resulting from exonuclease or restriction endonuclease activity are made compatible by the addition of complementary single-stranded nucleotide tails or by the addition of synthetic DNA linkers
      (1) Nucleotide tails are added to the 3′ ends of DNA molecules by an enzyme called terminal transferase
      (2) Foreign DNA and the vector are given different, complementary tails to facilitate the formation of base pairs (for example, polyadenosine tails are attached to one fragment and polythymidine tails to the other)
      (3) *Synthetic DNA linkers* are small, double-stranded DNA fragments that contain restriction sites; when they are ligated to blunt-ended DNA (vector and foreign DNA) and cut with the appropriate restriction enzyme, new sticky ends result

**C. Cloning vectors**

1. DNA sequences used as cloning vectors have several common characteristics
   a. Vectors are small, well-defined molecules that can be isolated from host cells easily
   b. They are self-replicating and contain an origin of replication
   c. They must contain restriction sites at which foreign DNA can be inserted
   d. They generally contain identifiable markers — genes that confer specific phenotypes (such as resistance to certain antibiotics)
2. In addition, cloning vectors commonly contain DNA sequences that confer specialized characteristics
   a. Vectors may contain the appropriate signals for the expression of the passenger DNA; these vectors are called *expression vectors*
   b. Vectors may contain both eukaryotic and prokaryotic origins of replication, allowing them to be shuttled from one organism to another; these are called *shuttle vectors*
3. Common vectors used in gene cloning include plasmids, bacteriophages, cosmids, and yeast artificial chromosomes
   a. Plasmid vectors are derived from naturally occurring prokaryotic genetic elements
      (1) Plasmids, such as pBR322 and pUC119, are circular, double-stranded DNA sequences

(2) Plasmid vectors can attain a high level of replication within the host cell (up to 500 copies per bacterium) and can be purified easily

(3) Disadvantages include a limit on the amount of passenger DNA that can be inserted and the reliance upon the less efficient process of transformation for introduction into the host genome

b. Bacteriophage vectors, such as lambda vectors, also are derived from naturally occurring genetic elements

(1) Lambda vectors contain a large, disposable region that can be replaced by a foreign DNA sequence

(2) Bacteriophage vectors generally can be used to clone larger fragments than plasmid vectors

(3) Bacteriophage vectors are introduced into host cells via the efficient process of transduction

(4) They are slightly more difficult to propagate and purify than plasmid vectors

c. ***Cosmids*** are hybrid vectors that contain sequences derived from both plasmids and lambda bacteriophages

(1) Cosmids contain a plasmid origin of replication (called an *ori*) and a lambda sequence (called *cos*) that allows for the packaging of vector and passenger DNA into phage particles for efficient transduction

(2) Cosmids are used to clone large DNA fragments — up to 50,000 base pairs (bps), or 50 kilobase pairs (kbps)

d. Yeast artificial chromosomes (YACs) are eukaryotic sequences that contain a yeast origin of replication, along with centromeric and telomeric sequences

(1) YACs are linear vectors that can carry extremely large DNA fragments (up to 1 million bps)

(2) When introduced into yeast cells, YACs behave like normal yeast chromosomes

(3) The cloning of large DNA fragments requires extremely delicate manipulations to avoid shearing the DNA

## D. Selection and bacterial transformation or transduction

1. Recombinant molecules are introduced into a host cell via transformation (with plasmid vectors) or transduction (with phage and cosmid vectors)

2. A marker on the vector, such as a gene conferring resistance to an antibiotic (for example, ampicillin or tetracycline), allows host cells that have acquired the vector to grow in an antibiotic-containing medium; cells that have not acquired the vector will not grow

3. Growth on an antibiotic-containing medium does not imply that the vector has incorporated a segment of passenger DNA; selection for host cells containing recombinant molecules requires an additional method of analysis

a. Typically, foreign DNA intentionally is inserted into a restriction site on the vector so that it disables a vector-borne gene

b. Selection for the disabled gene among those host cells known to have incorporated a vector permits the cells carrying foreign DNA to be identified

(1) Many vectors contain the *lac Z* gene, which is the gene required for the production of the enzyme β-galactosidase

(2) When a functional *lac Z* gene is present, β-galactosidase is produced and can be detected by growing bacterial cells on a chromogenic sub-

strate (a substrate that produces a colored product when the enzyme reacts with it)

(3) When the *lac Z* gene has been disrupted by insertion of foreign DNA, β-galactosidase is not present and cells will not produce a colored product on a chromogenic substrate

4. A common method for cloning a specific gene involves the cloning of an organism's entire genome in many small fragments — called a ***genomic library*** — and requires yet another level of selection for the detection of specific DNA sequences
   a. One method of gene detection requires the transfer of recombinant molecules to a nitrocellulose (or nylon) filter by placing the filter in contact with the bacterial cells or plaques growing on a petri dish
   b. Once the cells are ruptured and the recombinant molecules are covalently bound to the filter, the DNA is denatured to form single-stranded molecules
   c. The filter is placed into a solution that contains denatured, radiolabeled DNA that represents the gene of interest
      (1) The radiolabeled DNA, called a *probe,* may be a partially homologous (heterologous) DNA sequence previously isolated from another organism or it may be a synthetic DNA fragment that is based on the known protein sequence of the gene product
      (2) A fraction of the radiolabeled DNA will reanneal to the homologous DNA that is bound to the filter
      (3) After reannealing, the filter is gently washed and placed in contact with a sheet of film
      (4) The position of the reannealed probe on the filter, as revealed on the film, indicates the position of the bacterial cells or plaques on the petri dish that contain the gene of interest

**E. Complementary DNA cloning**

1. A genomic library contains all DNA sequences in a genome; some of the cloned molecules represent genes, while others represent the sequences that occur between genes
2. In order to clone only those DNA sequences that specify protein products (that is, the genes), geneticists must isolate ***messenger RNA (mRNA)***
   a. The mRNA molecule is a single-stranded molecule that is complementary to the DNA sequence that specifies it
   b. mRNA molecules arise only from those DNA sequences that specify proteins
3. Once mRNA is isolated, it can be converted into a complementary, single-stranded DNA molecule called ***complementary DNA (cDNA)***
   a. The enzyme ***reverse transcriptase*** catalyzes the formation of a DNA molecule from an mRNA template
   b. Once the single strand of cDNA is produced, the mRNA template can be degraded and replaced by a second DNA strand; this results in a double-stranded cDNA molecule
4. cDNA molecules can be cloned into vectors via the same method used for genomic DNA molecules
5. A collection of cDNA clones derived from a cell's pool of mRNA is called a *cDNA library*
6. Because a cDNA library represents only those DNA sequences that specify proteins, it is actually a subset of sequences found in a genomic library

7. DNA sequences that give rise to proteins are easier to find in a cDNA library than in a genomic library simply because the cDNA library is smaller
8. In addition, cDNA clones yield valuable information regarding the process of protein synthesis (see Chapter 11, Gene Expression, for details)

## IV. Analysis of Cloned Genes

### A. General information

1. After genes have been cloned, their structure and expression can be further investigated by a variety of methods
2. Virtually all of those methods depend upon the complementary nature of DNA and its ability to reanneal with homologous sequences after denaturation

### B. Southern blotting

1. One of the most useful techniques for analyzing cloned sequences was conceived by Edward Southern in 1975 and is called ***Southern blotting***
2. After digestion with restriction endonucleases, agarose-gel electrophoresis is used to separate genomic DNA fragments according to size
3. Then, the separated restriction fragments can be denatured and transferred to a nitrocellulose (or nylon) filter
   a. Denaturation is accomplished by soaking the gel in a dilute alkaline solution
   b. Restriction fragments are transferred to the filter by placing the gel on a buffer-soaked surface (such as a sponge), followed by the filter and a stack of absorbent towels
   c. The movement of the buffer (from the sponge through the gel and filter to the towels) results in the blotting of DNA molecules onto the surface of the filter while maintaining their exact positions relative to the other fragments
4. Once the DNA is covalently bound to the filter, it is exposed to a solution containing a denatured probe
   a. The probe will reanneal (hybridize) to the genomic DNA fragments on the filter that contains homologous sequences
   b. After washing the filter and exposing it to a sheet of film, the hybridized probe is visible on the film as a band of radioactivity
5. The relative position of the band reveals the size of the genomic DNA fragment that carries the specific sequence
6. This technique is useful for determining the number of sequences in a genome that are homologous to a cloned gene as well as for constructing restriction maps
7. The variability in sizes of restriction fragments — referred to as restriction fragment length polymorphisms (RFLPs) — is visualized on Southern blots and can be used as a genetic fingerprint (see *RFLP Analysis,* page 86)
   a. By using several different probes or by using one probe that hybridizes to many, distinct bands, geneticists can distinguish between individual genomes
   b. RFLPs can be mapped to specific chromosomal loci in a manner analogous that used for mapping visible mutations
   c. RFLPs are particularly useful for mapping disease-causing genes in humans
      (1) Cosegregation of an RFLP marker and a disease-causing gene indicates genetic linkage of the loci

## RFLP Analysis

Restriction fragment length polymorphism (RFLP) analysis is used to determine a genetic fingerprint. The darkened sequence on the restriction site map indicates the region of probe binding. The $A^2$ allele contains one additional restriction site, which results in the visualization of two bands on the Southern blot, as compared to allele $A^1$, which results in only one band. Southern blot analysis yields different banding patterns for each of three possible genotypes. Individuals may possess two $A^1$ alleles; two $A^2$ alleles; or one $A^1$ and one $A^2$ allele.

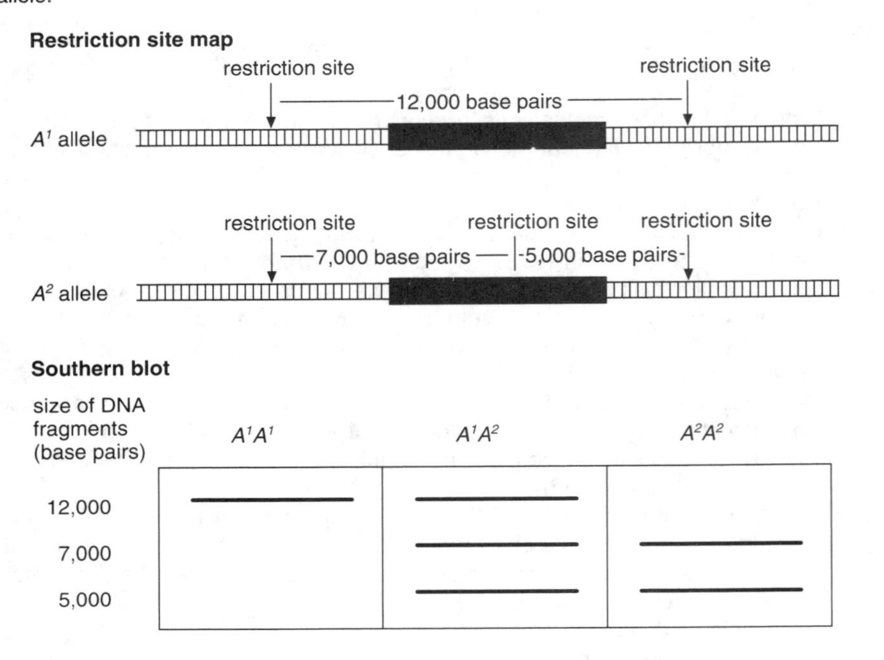


(2) RFLP analyses have been used to map the genes responsible for Duchenne muscular dystrophy (X chromosome), Huntington's disease (chromosome 4), cystic fibrosis (chromosome 7), neurofibromatosis (chromosome 17), and polycystic kidney disease (chromosome 16)

### C. Northern blotting

1. Whereas Southern blotting is useful in the analysis of gene structure, ***Northern blotting*** is useful in the analysis of gene function
2. The name Northern blotting is intended only to distinguish it from Southern blotting, the technique developed by Edward Southern
3. Northern blotting is similar to Southern blotting except that RNA molecules, rather than DNA restriction fragments, are separated by electrophoresis and transferred to a nitrocellulose filter
4. Hybridization with a specific probe will reveal whether the sequence is expressed, that is, whether the sequence is actively engaged in protein synthesis

a. Comparison of Northern blots using RNA from different tissues, or from different stages of the cell life cycle, reveals where and when specific genes are turned on

b. The relative amount of hybridized RNA is an indication of the level of gene expression in a particular cell or tissue

**D. Polymerase chain reaction**

1. Molecular analyses of gene structure and function traditionally have required the cloning of specific sequences in order to obtain enough DNA for study
2. A more recent technique, however, results in the amplification of specific DNA sequences without cloning
3. The ***polymerase chain reaction (PCR)*** involves the in vitro replication of any sequence for which the terminal nucleotide sequences are known
   a. Knowledge of the terminal nucleotide sequence allows for the synthesis of short primers that contain 10 to 20 nucleotides and flank the sequence to be amplified
   b. Denaturation of the DNA to be studied facilitates the hybridization of the primers to each end of the region to be amplified
   c. In the presence of deoxyribonucleotides, DNA polymerase extends the primers from the 3′ end
   d. Many cycles of denaturation, reannealing, and primer extension ultimately result in billions of replication products (see *Polymerase Chain Reaction,* page 88)
   e. PCR amplification can be used to detect DNA polymorphisms and to provide DNA for restriction mapping or for use as radiolabeled probes

**E. DNA sequencing**

1. The ability to determine the exact nucleotide sequence of a DNA fragment led to a fuller understanding of gene structure and function
   a. For example, the molecular basis of specific gene mutations can be investigated
   b. DNA sequencing has revealed regulatory regions that are common to all genes
   c. The comparison of DNA sequencing among different species has provided an estimate of molecular evolution
2. DNA sequencing techniques depend upon the ability to generate a collection of DNA fragments whose sizes differ from one another by a single nucleotide
3. Two methods have been widely used for DNA sequencing: Alan Maxam and Walter Gilbert's base-destruction method (developed in 1977) and Frederick Sanger's dideoxy method (also developed in 1977)
   a. The base-destruction method uses chemical reagents to randomly destroy specific bases in a single-stranded DNA fragment that have radioactive label attached to one end
      (1) Four reactions are utilized: the random destruction of guanine (G) only; the random destruction of adenine and guanine (A and G); the random destruction of thymine and cytosine (T and C); and the random destruction of cytosine (C) only
      (2) In each reaction, DNA fragments of different lengths are produced, according to the site of base destruction; this results in DNA breakage

## Polymerase Chain Reaction

The polymerase chain reaction begins when a double-stranded DNA molecule that contains the target sequence is denatured and reannealed to primers. Primers are constructed so that they are complementary to the 3′ regions of the target sequence; Taq polymerase extends these primers. When the sequences are denatured again, the old and new target DNA strands will anneal to primers. Taq polymerase is used again to extend the primers. The process continues, resulting in amplification of target sequences. The diagram below illustrates this process.

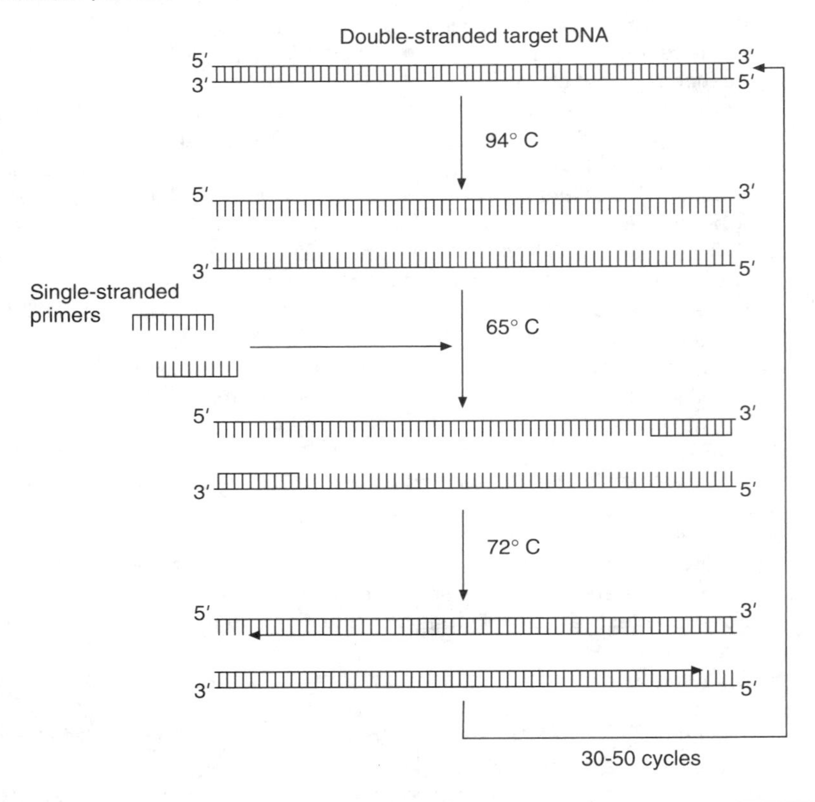


(3) The DNA fragments are subjected to electrophoresis, and four columns containing discrete bands that represent DNA fragments of a specific length are visible after film exposure

(4) The difference in DNA length between bands is a single nucleotide; the specific nucleotide can be deduced by observing the reaction in which the band was produced

b. The dideoxy method relies on the random termination of DNA synthesis caused by the incorporation of four dideoxynucleoside triphosphates

(1) A small primer is first annealed to the end of a single-stranded DNA template

(2) The primer is extended by DNA polymerase in the presence of all four dideoxynucleoside triphosphates

(3) This reaction is repeated four times, but each contains a small quantity of one dideoxynucleoside triphosphate (dideoxy A, dideoxy T, dideoxy C, or dideoxy G)
(4) DNA polymerization will terminate after the incorporation of a dideoxynucleotide because the lack of a hydroxyl group prevents the binding of the next nucleotide
(5) Each reaction thus contains a mixture of DNA fragments; each fragment in a particular reaction is terminated at a specific nucleotide
(6) Electrophoresis separates these molecules into discrete bands; the difference in DNA length between bands is a single, known nucleotide

## Study Activities

1. You have isolated a linear fragment of DNA. After cutting the fragment with *Eco*RI and performing electrophoresis using an agarose gel and autoradiography, you observe two bands, one representing 5-kbp molecules and one representing 9-kbp molecules. When the procedure is repeated with *Eco*RV, you observe a 2-kbp band and a 12-kbp band. When both enzymes are used in the same reaction, three bands result: a 2-kbp band, a 5-kbp band, and a 7-kbp band. Draw a restriction site map for this DNA fragment and indicate the distance between restriction sites.
2. Draw a flow diagram that depicts genomic cloning.
3. Make a table that summarizes the characteristics, advantages, and disadvantages of various cloning vectors.
4. You have just constructed a human genomic library and would like to isolate the gene for a specific enzyme that you are studying. You have already cloned a highly homologous mouse gene. How might you select your human clone?
5. After isolating the human clone described in question 4, you would like to determine whether the gene is active in the liver. How can you do this?
6. Compile a list of applications for which you feel that genomic fingerprinting using RFLP or PCR technology may be beneficial.
7. If a primer 3′-ATGTGG-5′ containing 6 bps is annealed to the strand 5′-AAGTTCGCATGGGCAGATCGTACACC-3′ and is used in four dideoxy sequencing reactions, what fragments will result from each reaction?

# 11

## Gene Expression

### Objectives

After studying this chapter, the reader should be able to:

- Describe all possible directions of genetic information transfer, including the processes of replication, transcription, translation, reverse transcription, and ribonucleic acid (RNA) self-replication.
- List and describe the molecules required for transcription and translation.
- Explain the mechanisms of initiation, protein elongation, and termination during translation.
- Determine the amino acid sequence encoded by a specific deoxyribonucleic acid (DNA) sequence.
- List and describe the differences between prokaryotic and eukaryotic protein synthesis.

## I. Overview of Protein Synthesis

**A. General information**

1. Remember that genes specify proteins via an mRNA intermediate
2. The production of RNA from a DNA template is called ***transcription***
   a. We say that DNA, or a gene, is transcribed into RNA
   b. Several types of RNA intermediates are produced, including messenger RNA (mRNA), transfer RNA (tRNA), ribosomal RNA (rRNA), and ***small nuclear RNA (snRNA)***
3. The production of protein from an RNA intermediate is called ***translation***

**B. Colinearity of genes and proteins**

1. Many details of protein synthesis were discovered during the 1960s, several years after Watson and Crick proposed the double helix model of DNA
2. The research of Charles Yanofski in 1967 provided evidence that the linear sequence of nucleotides in a gene determined the linear sequence of amino acids in the corresponding protein
   a. The order of mutational sites in the *trypA* gene of *Escherichia coli* was found to be the same as the order of corresponding amino acid changes in the resulting protein
   b. Genes and proteins therefore are colinear
3. The exact correspondence between a gene and a protein was later demonstrated when DNA sequencing techniques were developed in the 1970s by Walter Gilbert, Alan Maxam, and Frederick Sanger

## II. Transcription

### A. General information

1. Transcription of DNA into RNA is based on the complementarity of bases
2. The production of RNA from a DNA template is similar to the production of DNA from a DNA template (replication), with some notable exceptions
    a. First, RNA synthesis involves only one strand of a double-stranded DNA molecule
        (1) The strand of DNA that is used as a template is called the *sense strand;* the strand of DNA that does not participate in transcription is called the *antisense strand*
        (2) Although only one strand of DNA is transcribed, the same strand is not necessarily transcribed for all the genes that occur on a particular DNA molecule
    b. Second, RNA contains the sugar ribose, not deoxyribose (which is in DNA)
    c. Third, RNA contains the pyrimidine uracil where thymine exists in the DNA molecule

### B. Enzymology

1. RNA synthesis is catalyzed by ***RNA polymerase*** in the presence of a DNA template, magnesium ions, and nucleoside triphosphates
    a. Most prokaryotic cells contain a one type of RNA polymerase
    b. Eukaryotic cells contain three types of RNA polymerases
        (1) RNA polymerase I catalyzes the formation of rRNA
        (2) RNA polymerase II catalyzes the formation of mRNA
        (3) RNA polymerase III catalyzes the formation of tRNA, snRNA, and some rRNA
2. RNA polymerase does not require a primer to initiate a new RNA strand
3. RNA polymerase does not have a proofreading function, as DNA polymerase does
4. RNA polymerase is a complex enzyme that contains multiple subunits
    a. The RNA polymerase of *E. coli* is the most extensively studied of all RNA polymerases
    b. One of the *E. coli* RNA polymerase subunits, the ***sigma factor,*** is required for transcription initiation
    c. The other subunits, called the core enzyme, are required for elongation of the nascent RNA
    d. The complete enzyme is called the ***holoenzyme***

### C. Initiation of transcription

1. The sigma factor of the prokaryotic holoenzyme recognizes a specific DNA sequence, called the ***promoter,*** which occurs at the beginning of a gene; in eukaryotes, RNA polymerase recognizes the promoter with the assistance of ***transcription factors***
    a. In prokaryotes, promoters generally contain two less variable (most conserved) sequences
        (1) The first sequence is centered approximately 35 nucleotides from the transcription start site and is called the *–35 region;* the second sequence is centered approximately 10 nucleotides from the transcription start site and is called the *–10 region* (or ***Pribnow box***)

(2) By comparing promoter sequences from many different genes, a generalized or *consensus* sequence can be recognized
(3) The –35 consensus sequence is 5′-TTGACA-3′
(4) The –10 consensus sequence is 5′-TATAAT-3′
b. In eukaryotes, promoters are most conserved at two regions
(1) The first sequence is centered approximately 75 nucleotides from the transcription start site and is called the CAAT box; the second sequence is centered approximately 25 nucleotides from the transcription start site and is called the TATA box or ***Hogness box***
(2) The consensus sequence for the CAAT box is 5′-GGCCAATCT-3′
(3) The consensus sequence for the TATA box is 5′-TATAAAA-3′
2. After it recognizes the promoter sequence, the RNA polymerase of *E. coli* binds tightly to the DNA and denatures (unwinds) approximately 17 nucleotide pairs in the –10 region

### D. RNA elongation during transcription

1. Shortly after initiation, the sigma factor dissociates from the core enzyme
2. The RNA polymerase core enzyme continues to move along the DNA double helix, catalyzing RNA synthesis in the 5′ → 3′ direction; new nucleotides are added to the 3′-OH end of the nascent RNA molecule
3. Unwinding of the DNA helix in one region results in ***supercoiling*** of the DNA in other regions
4. Supercoiling is created upstream and downstream of the transcription site and is alleviated by the enzyme topoisomerase

### E. Termination of transcription

1. Elongation of the RNA molecule continues until the RNA polymerase encounters a terminator sequence in the DNA
2. Two types of prokaryotic terminators exist: rho-dependent terminators and rho-independent terminators
a. As the name implies, rho-dependent termination depends upon a protein factor called ***rho***
(1) Geneticists believe that rho binds to the RNA molecule and causes an unwinding of the RNA-DNA complex
(2) After unwinding occurs, the DNA, RNA, and rho factor dissociate and transcription stops
b. Rho-independent termination depends upon the formation of a ***stem-loop structure*** in the RNA molecule
(1) A stem-loop structure results from the folding back of the RNA molecule onto itself in a region with inverted, repeated sequences
(2) Stem-loop structures apparently cause the RNA polymerase to stall
(3) When the RNA polymerase stalls, an adjacent string of uracil-containing nucleotides facilitates the destabilization of the RNA-DNA complex because of less hydrogen bonding

## III. RNA Molecules

### A. General information

1. As described above, RNA molecules include mRNA, rRNA, tRNA, and snRNA

2. Many RNA molecules must be processed before they can become biologically active
   a. Initial RNA transcripts are called ***precursor RNAs (pre-RNAs)*** and represent the primary transcripts
   b. Formation of mature RNA molecules may involve the addition, deletion, or modification of nucleotides

**B. Messenger RNA and small nuclear RNA**

1. Messenger RNA molecules have three distinct regions: a 5′ leader sequence, a protein-coding sequence, and a 3′ trailing sequence
   a. The leader sequence is important for initiating translation, but none of the bases in this sequence codes for amino acids
   b. The protein-coding sequence is the region that directly determines the primary structure of the protein
   c. The trailing sequence does not code directly for amino acids, but it may play a role in determining the stability of the mRNA molecule code for amino acids
2. Prokaryotic mRNA molecules seldom are modified, whereas eukaryotic mRNA molecules are extensively processed
   a. Posttranscriptional modification of eukaryotic pre-mRNAs includes addition of a 5′ cap and a 3′ tail
      (1) The 5′ cap consists of a methylated guanine nucleotide attached to the 5′ terminal nucleotide by a 5′-5′ linkage, as opposed to a 5′-3′ linkage
      (2) The 3′ tail consists of a string of 50 to 250 adenine nucleotides; this is called a poly-A tail
      (3) Neither the 5′ cap or the 3′ tail is represented in the DNA template
      (4) The cap and tail structures are required for initiation of translation and increased mRNA stability, respectively
   b. Posttranscriptional modification of eukaryotic pre-mRNAs also may include the removal of noncoding sequences from the protein-coding region of the mRNA
      (1) These noncoding sequences are called ***introns,*** or intervening sequences
      (2) Introns are excised prior to mRNA transport into the cytoplasm
      (3) Protein-coding sequences that remain after intron removal are called ***exons,*** or expressed sequences
      (4) The number and extent of introns can be visualized by reannealing denatured DNA and the corresponding mature mRNA
         (a) An RNA-DNA hybrid will displace the non-template (nonsense) DNA strand
         (b) Regions of the DNA template that encode introns will not have a complementary region in the mature mRNA and will therefore be visible as single-stranded loops
      (5) mRNA splicing requires specific nucleotide sequences at the 5′ and 3′ boundaries of the intron, which are called splice junctions
         (a) The first event in intron removal is mRNA cleavage at the 5′ boundary of the intron
         (b) The unbound 5′ end of the intron becomes bound to a specific site within the intron — the branch-point sequence — by a 2′-5′ bond

(c) The looped intron now resembles a lariat and is called a ***lariat structure***

(d) Cleavage of the mRNA at the 3′ end of the intron, coupled with simultaneous ligation of flanking exons, results in mature mRNA and release of the lariat structure

(6) Intron splicing is mediated by a complex of snRNAs

(a) Six principal snRNAs (labeled U1, U2, U3, U4, U5, and U6) exist in the cell nucleus

(b) snRNAs are associated with proteins; snRNA-protein complexes are called *small nuclear ribonucleoprotein particles (snRNPs)*

(c) snRNPs are the components of the ***splicesome,*** the complex of RNA and protein that catalyzes excision of introns

(7) Some intron splicing is catalyzed by the pre-mRNA molecule itself; the RNA that participates in this process is called self-splicing RNA

(a) For example, the intron of the *Tetrahymena* rRNA precursor (called a group I intron) catalyzes its own excision

(b) RNA molecules that have enzymatic properties are called ***ribozymes***

## C. Transfer RNA

1. Transfer RNA molecules transport amino acids to the site of protein synthesis and then translate the base sequence of mRNA into the amino acid sequence of a protein
2. Each amino acid is carried by a specific tRNA
   a. Amino acids are attached to the appropriate tRNA molecules by ***aminoacyl-tRNA synthetases***
   b. Although separate synthetases exist for each amino acid, all synthetases need adenosine triphosphate (ATP) to link a tRNA molecule with an amino acid
3. Like many other RNA molecules, the primary transcripts of tRNA genes can be extensively modified
4. A mature tRNA molecule can assume a complex secondary structure that involves folding of the molecule and forming intramolecular base pairs
   a. The two-dimensional shape of a mature tRNA molecule resembles a cloverleaf structure; molecules contain a long stem (containing the 5′ and 3′ ends of the molecule) and three primary loops (each with a small stem)
   b. The long stem (the acceptor stem) is the site of amino acid attachment
   c. One loop (the T-loop, or loop IV) is responsible for recognizing the ribosome, which is the site of protein synthesis
   d. One loop (the anticodon loop, or loop II) is responsible for forming base pairs with the mRNA during the translation process
      (1) The forming of base pairs between a tRNA and an mRNA involves three bases on each molecule
      (2) The three participating bases on the mRNA are collectively called the ***codon***
      (3) The three participating bases on the tRNA are collectively called the ***anticodon***
   e. Another loop (the D-loop) may be involved in recognition of the appropriate synthetase molecule

### D. Ribosomal RNA

1. As their name implies, rRNAs are components of the ribosome, the site of protein synthesis
   a. Approximately two-thirds of a ribosome's mass consists of rRNA; the remaining one-third represents ribosomal proteins
   b. Ribosomes contain two subunits — one large and one small
      (1) The prokaryotic subunits are 50 Svedberg units (S) and 30S in size; the entire prokaryotic ribosome is 70S
      (2) The eukaryotic subunits are 60S and 40S in size; the entire eukaryotic ribosome is 80S
2. The prokaryotic ribosome contains three different rRNA molecules: 23S and 5S rRNAs in the large subunit and 16S rRNA in the small subunit
3. The eukaryotic ribosome contains four different rRNA molecules: 28S, 5.8S, and 5S rRNAs in the large subunit and 18S rRNA in the small subunit
4. The genes that encode different rRNA molecules are called rDNA; in prokaryotes and eukaryotes, they are arranged next to one another and are transcribed as a single pre-rRNA
   a. In prokaryotes, the genes encoding the 16S, 23S, and 5S rRNAs are arranged, in the order given, into a single transcription unit
   b. In eukaryotes, the genes encoding the 18S, 5.8S, and 28S rRNA are arranged, in the order given, into a single transcription unit; the 5S genes are not transcribed as part of this unit
   c. Pre-rRNAs are cleaved at specific sites by ***RNases,*** enzymes that catalyze cleavage of RNA molecules

## IV. Translation

### A. General information

1. Once a mature mRNA is produced and transported into the cytoplasm (as in the case of eukaryotes), the sequence of nucleotide bases is translated into a sequence of amino acids
2. The translation process requires the interaction of mRNA, tRNA, and rRNA, along with other components
3. For the purpose of discussion, translation can be divided into three stages: initiation, elongation, and termination

### B. The genetic code

1. The 20 known amino acids are specified by the sequence of bases in an mRNA molecule
   a. If we first assume that each base (adenine [A], cytosine [C], guanine [G], and uracil [U]) specifies one amino acid, then only 4 amino acids can be encoded
   b. Likewise, if two bases specify an amino acid (for example, AU or GC), only 16 amino acids can be encoded
   c. If three bases specify an amino acid (AUG or CUU, for example), all 20 amino acids can be specified by a unique three-base combination
2. The genetic code is a triplet code; each amino acid of a growing protein is specified by a combination of three bases called a codon (see *The Genetic Code,* page 96)

## The Genetic Code

The three-base combinations shown in the chart below are the messenger ribonucleic acid (mRNA) codons needed for replication. Each codon — with the exception of the three terminator (stop) codons — encodes an amino acid. The deoxyribonucleic acid (DNA) templates and the transfer RNA (tRNA) anticodons for each codon would necessarily be the complementary sequence. The abbreviations for the amino acids are listed below the chart.

| First position (5′ end) | Second position U | C | A | G | Third position (3′ end) |
|---|---|---|---|---|---|
| U | UUU Phe | UCU Ser | UAU Tyr | UGU Cys | U |
| | UUC Phe | UCC Ser | UAC Tyr | UGC Cys | C |
| | UUA Leu | UCA Ser | UAA Stop | UGA Stop | A |
| | UUG Leu | UCG Ser | UAG Stop | UGG Trp | G |
| C | CUU Leu | CCU Pro | CAU His | CGU Arg | U |
| | CUC Leu | CCC Pro | CAC His | CGC Arg | C |
| | CUA Leu | CCA Pro | CAA Gln | CGA Arg | A |
| | CUG Leu | CCG Pro | CAG Gln | CGG Arg | G |
| A | AUU Ile | ACU Thr | AAU Asn | AGU Ser | U |
| | AUC Ile | ACC Thr | AAC Asn | AGC Ser | C |
| | AUA Ile | ACA Thr | AAA Lys | AGA Arg | A |
| | AUG Met | ACG Thr | AAG Lys | AGG Arg | G |
| G | GUU Val | GCU Ala | GGU Asp | GGU Gly | U |
| | GUC Val | GCC Ala | GGC Asp | GGC Gly | C |
| | GUA Val | GCA Ala | GGA Glu | GGA Gly | A |
| | GUG Val | GCG Ala | GGG Glu | GGG Gly | G |

**Abbreviations**

Alanine — Ala
Arginine — Arg
Asparagine — Asn
Aspartic acid — Asp
Cysteine — Cys
Glutamic acid — Glu
Glutamine — Gln
Glycine — Gly
Histidine — His
Isoleucine — Ile
Leucine — Leu
Lysine — Lys
Methionine — Met
Phenylalanine — Phe
Proline — Pro
Serine — Ser
Threonine — Thr
Tryptophan — Trp
Tyrosine — Tyr
Valine — Val

3. Codons are read sequentially, without skipping any bases, and in a nonoverlapping manner
4. Because there are 64 possible codons, some amino acids are encoded by more than one codon
5. The genetic code is considered *degenerate* because more than one codon can specify a single amino acid
6. One specific codon (AUG), called the *start codon,* indicates the site on the mRNA at which translation is to start; the codons UAG, UGA, and UAA indicate the termination of translation and are called *termination codons*, or stop codons
   a. Besides indicating the translation start site, the start codon codes for the amino acid methionine
   b. The termination codons do not specify the incorporation of any amino acids
7. With minor exceptions, the genetic code is the same for all organisms; therefore, it is a universal code

### C. Initiation of translation

1. In prokaryotes, initiation of translation begins with the binding of the 30S ribosomal subunit to the mRNA
   a. An initiation factor, IF3, assists in the binding
   b. The bound 30S ribosomal subunit is not attached to the 50S subunit at the time of initiation, but exists in a free form
2. A second initiation factor, IF2, then binds to both guanosine triphosphate (GTP) and the initiator tRNA and directs the tRNA to the mRNA-30S complex
   a. The initiator tRNA contains the anticodon, which is complementary to the start codon
   b. The initiator tRNA carries a modified methionine molecule (N-formylmethionine, or fMet) rather than normal methionine
3. The final step of initiation is the union of the ribosomal subunits, which is driven by the cleavage of the GTP molecule
4. The mRNA then is positioned so that translation will begin at the start codon, AUG
   a. The prokaryotic initiator tRNA also can recognize GUG and, rarely, UUG as start codons
   b. In prokaryotes, the start codon is distinguished from other methionine codons by its proximity to a specific nucleotide sequence occurring near the 5′ end of the coding region, called the Shine-Delgarno sequence
      (1) The Shine-Delgarno sequence facilitates binding of the mRNA to the 16S rRNA component of the small ribosomal subunit so that the start codon is in appropriate alignment with the ribosome
      (2) Eukaryotic mRNAs do not contain Shine-Delgarno sequences
5. The assembled ribosome contains two tRNA-binding sites, called the A-site and the P-site
   a. The A-site is the binding site for aminoacyl-tRNAs (the tRNAs carrying amino acids)
   b. The P-site is the site at which the initiator tRNA binds during initiation and the peptidyl-tRNA (the tRNA carrying the growing peptide) is bound during elongation

### D. Protein elongation during translation

1. When the initiator tRNA is bound to the P-site during the initiation phase, the anticodon bases are paired with the bases in the mRNA start codon
2. The second codon of the mRNA is aligned so that it can form base pairs with the anticodon of an aminoacyl-tRNA bound at the A-site (if the bases are complementary), thus starting the elongation phase of protein synthesis
   a. In prokaryotes, aminoacyl-tRNAs are transported to the ribosome's A-site after forming a complex with an ***elongation factor,*** called EF-Tu, and a molecule of GTP
   b. When the aminoacyl-tRNA is positioned properly, the GTP molecule is hydrolyzed and a complex consisting of guanosine diphosphate (GDP) and EF-Tu is released
   c. The GDP/EF-Tu complex is "recharged" into a new GTP/EF-Tu complex through the action of a second elongation factor, called EF-Ts
3. If the codon-anticodon pairing is correct, a peptide bond is formed between the amino acid on the tRNA located at the A-site and the amino acid on the tRNA located at the P-site

a. For some codon-anticodon combinations, peptide bonds can form although the base-pairing is not exact
(1) Some tRNAs can form base pairs with more than one codon because of "sloppy" bond formation at one end of the codon
(2) This phenomenon — called *wobble* — accounts for some of the degeneracy in the genetic code
b. The peptide bond is catalyzed by the enzyme ***peptidyl transferase***
c. Bond formation involves breakage of the bond between the amino acid and the tRNA at the P-site, followed by peptide bond formation
4. The tRNA located at the P-site no longer is carrying an amino acid, whereas the tRNA at the A-site (now called a peptidyl-tRNA) contains the growing peptide
5. After peptide bond formation, the ribosome moves one codon down the mRNA
a. This translocation (movement) of the ribosome is facilitated by yet another elongation factor, EF-G, and another molecule of GTP
b. Translocation results in the ejection of the uncharged tRNA at the P-site and the transfer of the peptidyl-tRNA from the A-site to the P-site
c. Once it is empty, the A-site binds to the next appropriate aminoacyl-tRNA
6. As the process continues, each incoming aminoacyl-tRNA binds at the A-site, thereby facilitating peptide bond formation and translocation of the ribosome to the next codon
a. The peptide is synthesized from its amino end (N-terminus) to its carboxyl end (C-terminus)
b. Translation occurs in a $5' \rightarrow 3'$ direction, so that the ribosome always moves from the 5′ end of the mRNA to the 3′ end

### E. Termination of translation

1. The elongation phase continues until a termination codon is encountered at the A-site
2. Because there are no tRNAs that form base pairs with the stop codons, protein synthesis is terminated
3. The action of release factors facilitates the release of the newly formed protein
a. Prokaryotes contain three release factors (RF-1, RF-2, and RF-3), whereas eukaryotes contain only one (eRF)
b. Release factors recognize the termination codons when they are present at the A-site and cause the protein synthesis apparatus to dissociate

## V. Comparison of Prokaryotic and Eukaryotic Protein Synthesis

### A. General information

1. Several differences between prokaryotic and eukaryotic protein synthesis already have been introduced
a. The number and nature of RNA polymerases differ
b. Promoter consensus sequences differ
c. Eukaryotic mRNA molecules are modified extensively at the 5′ and 3′ ends, whereas prokaryotic mRNAs are not

d. Eukaryotic mRNAs generally contain one or more introns
e. The number and size of ribosomal components in eukaryotes and prokaryotes differ
f. Eukaryotic mRNAs do not contain Shine-Delgarno sequences

2. In addition, prokaryotic mRNAs generally are polycistronic — meaning that a single mRNA molecule contains more than one protein-coding region — whereas eukaryotic mRNAs are monocistronic

### B. Simultaneous transcription and translation in prokaryotes

1. Because prokaryotes lack a nuclear membrane, translation begins before transcription ends
2. Because mRNA molecules are synthesized from the 5′ end to the 3′ end, ribosomes can attach to the nascent mRNA and begin translation as soon as the RNA is produced

## VI. Variations in Information Transfer

### A. General information

1. The flow of genetic information from DNA to DNA (replication) and from DNA to RNA to protein (transcription and translation) has been referred to as the central dogma of genetics
2. Recent discoveries have led geneticists to conclude that genetic information can be transferred in additional ways

### B. Reverse transcription

1. RNA tumor viruses and the human immunodeficiency virus (HIV) can direct the synthesis of DNA from an RNA template
2. RNA-dependent DNA synthesis, called reverse transcription, is catalyzed by the enzyme reverse transcriptase
3. RNA viruses that direct RNA-dependent DNA synthesis are called ***retroviruses*** (see Chapter 17, Transposable Elements, Retroviruses, and Oncogenes)

### C. RNA self-replication

1. Some simple RNA phages can synthesize new RNA from an RNA template in a process called RNA self-replication
2. The enzyme ***RNA replicase*** catalyzes RNA self-replication

## Study Activities

1. Make a diagram illustrating all possible directions of genetic information transfer. Include the name of each process.
2. Match the name of the enzyme on the left with its function on the right:

| | |
|---|---|
| RNA polymerase | RNA-dependent DNA synthesis |
| RNA replicase | cleavage of RNA phosphodiester bonds |
| aminoacyl-tRNA synthetase | DNA-dependent RNA synthesis |
| RNase | RNA self-replication |
| peptidyl transferase | tRNA charging |
| reverse transcriptase | formation of peptide bonds |

3. Define the following terms: sigma factor, promoter, transcription factor, intron, ribozyme, and elongation factor.
4. List the roles of each type of RNA molecule.
5. The double-stranded DNA sequence below is a portion of a gene that encodes a hypothetical, 6-amino acid protein. Which of the strands, when transcribed, will produce an mRNA molecule with appropriately spaced start and stop codons? Using the information in *The Genetic Code* (page 96), determine the amino acid sequence of this protein.

   5′-CTAATGGTCTATTACTAACAT-3′
   3′-GATTACCAGATAATGATTGTA-5′
6. Consider the imaginary protein sequence methionine-leucine-methionine-tryptophan-serine-threonine. How many different mRNA sequences could produce this protein?
7. Draw a diagram that illustrates the movement of tRNA molecules in and out of the A-site and the P-site of the ribosome during protein elongation.

# 12

# Control of Gene Expression in Prokaryotes

## Objectives

After studying this chapter, the reader should be able to:

- Describe the various structural and regulatory elements of an operon.
- Differentiate between inducible and repressible operon systems.
- Distinguish between negative and positive transcriptional control.
- Describe the concept of catabolite repression.
- Discuss the concept of attenuation.
- Explain the complex interaction of regulatory mechanisms that controls gene expression.

## I. Concept of the Operon

### A. General information

1. Prokaryotic cells, such as those of *Escherichia coli,* do not require the same gene products in all environments
2. Some proteins are present in all cells at all times, and there is little variation in the production of these proteins
    a. This type of gene expression is called ***constitutive gene expression***
    b. The production of other proteins is more tightly regulated
3. The primary mechanisms that control the extent of gene expression involve regulation of transcription
4. Because prokaryotic messenger ribonucleic acid (mRNA) molecules are relatively short-lived (with a half-life of approximately 2 minutes), a change in the transcription rate quickly alters the rate of protein production
5. Because prokaryotic genes typically are transcribed into a polycistronic mRNA, transcriptional regulation will simultaneously control the production of several different proteins
6. A unit of coordinate gene expression is called an ***operon***

### B. Induction vs. repression

1. Proteins that are required only by cells in certain environments are produced by operons that generally are transcriptionally silent but can be rapidly induced; such operons are labeled *inducible*
2. Other proteins are produced by operons that generally are transcriptionally active; such operons are repressed when they are no longer needed and are labeled *repressible*

3. In short, inducible operons are transcriptionally active when a specific chemical is present, whereas repressible operons are transcriptionally inactive when a specific chemical is present

## II. The *Lac* Operon: An Inducible System

### A. General information

1. *E. coli* cells use lactose as a source of carbon by synthesizing enzymes that transport lactose into the cell and cleave the disaccharide to yield glucose and galactose
   a. Lactose permease is responsible for the active transport of lactose across the cell membrane
   b. ß-galactosidase is responsible for cleavage of the disaccharide
   c. Transacetylase is an enzyme whose role in the metabolism of lactose is not understood fully
   d. These three enzymes are encoded by genes that comprise the lactose, or *lac,* operon
2. In the absence of lactose, the enzymes that catalyze lactose degradation are not needed and their production is transcriptionally repressed
   a. A specific ***repressor protein,*** the lac repressor, can bind to the *lac* operon and effectively prevent RNA polymerase from transcribing the polycistronic mRNA
   b. Lactose and lactose derivatives can stimulate the production of the appropriate enzymes by interacting with the *lac* repressor; accordingly, they are called inducers

### B. Structure of the *lac* operon

1. The enzymes ß-galactosidase, lactose permease, and transacetylase are encoded by genes *lacZ, lacY,* and *lacA,* respectively (for more information, see *The* Lac *Operon*)
2. The enzyme-encoding genes, called ***structural genes,*** are adjacent to one another and are transcribed as a single unit
3. The single transcriptional unit is regulated by a single promoter and an adjacent sequence called an ***operator***
   a. The operator is the binding site for the *lac* repressor
   b. Because of the proximity of the operator and promoter, binding of the *lac* repressor at the operator prevents RNA polymerase from binding at the promoter
4. The *lac* repressor is encoded by a ***regulatory gene*** called the *I* gene, which has its own promoter and therefore is independently transcribed

### C. Induction

1. In the absence of lactose, the *lac* repressor binds to the operator of the *lac* operon and transcription is inhibited
2. When present, lactose molecules bind to the *lac* repressor at an allosteric site (one that is separate from the operator-binding site) and change the conformation of the repressor molecule

### The *Lac* Operon

The regulatory gene *(I)* encodes the repressor protein, which binds to the operator *(O)* in the absence of lactose. When present, lactose binds to the repressor and causes it to lose its operator-binding ability. Ribonucleic acid (RNA) polymerase then transcribes the structural genes (*Z, Y,* and *A*) as a single unit.

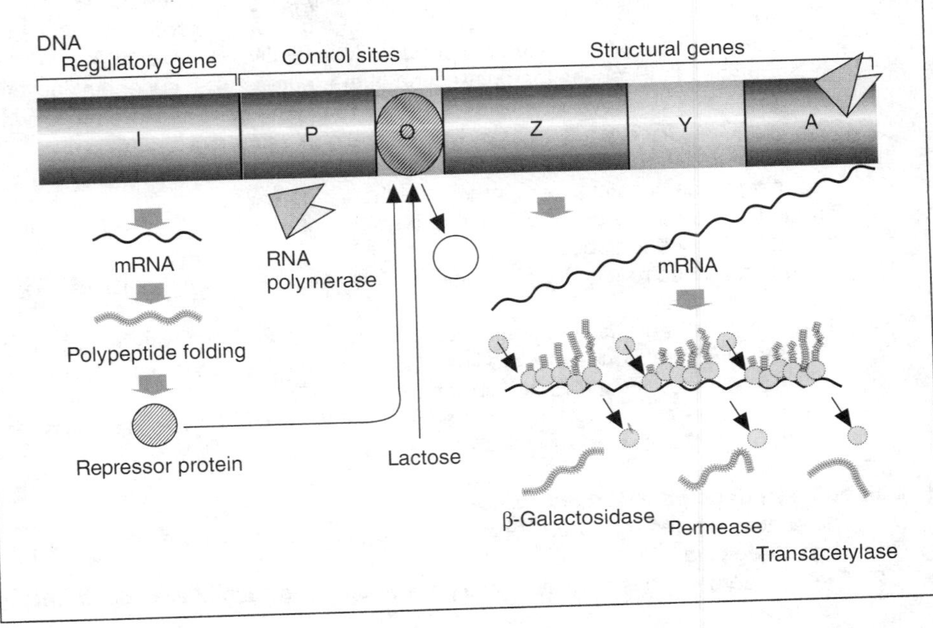


3. The conformational change invoked by the binding of lactose prevents the repressor from binding to the operator; thus, the repressor is removed from the operon and RNA polymerase can initiate transcription
4. Therefore, the enzymes required for lactose metabolism are produced only when lactose is present

**D. Effects of mutation**

1. The mechanism of *lac* operon induction was elucidated by François Jacob and Jacques Monod in 1961 when they observed bacterial strains that carried mutations in the *lac* genes
2. Mutations in the structural genes affect only the genes in which the mutation is located
   a. A mutation in *lacZ* (represented by $lacZ^-$) results in a loss of ß-galactosidase activity; lactose permease and transacetylase activities are unaffected
   b. Likewise, $lacY^-$ and $lacA^-$ mutations result in the loss of function of their respective enzymes
   c. An exception to this rule was observed with nonsense mutations — that is, the creation of a stop codon in the protein-coding region
      (1) A nonsense mutation in lacZ affects the production of ß-galactosidase a well as the other two enzymes

(2) This type of mutation, called a *polar mutation,* influences all genes downstream from the mutation site because the ribosome terminates translation and falls off the polycistronic message when it encounters the inappropriate stop codon

3. Mutations in the *I* gene affect enzyme production, not enzyme activity
   a. A mutation that causes the destruction of the repressor's operator binding site (called $I^-$) results in the production of *lac* operon-encoded enzymes in the presence or absence of an inducer (constitutive expression)
   b. A mutation that causes the destruction of the repressor's inducer binding site (called $I^s$) destroys the ability of lactose to induce gene expression; accordingly, the enzymes are not produced in any environment
4. A mutation in the operator (called $O^c$) destroys the repressor's ability to bind, resulting in the constitutive expression of the *lac* genes
5. Partial diploids can be constructed to determine dominance relationships for the mutations described above
   a. Structural and regulatory genes can be introduced into *E. coli* cells via ***sexduction***
   b. $O^c$ mutations are *cis*-dominant because they are dominant with respect to the structural genes that are linked to the mutant operator; operator mutations do not affect unlinked structural genes
   c. $I^s$ mutations are *trans*-dominant because the repressor protein is a diffusible product that can affect unlinked operons

### E. Catabolite repression of the *lac* operon

1. The control exerted by the *lac* repressor is negative; in other words, the repressor has a negative effect on gene expression
2. The *lac* operon also is regulated by a positive control system, called ***catabolite repression***
   a. Catabolite repression ensures that lactose will be used as an energy source only when glucose is absent
   b. If glucose is absent, cyclic adenosine monophosphate (cAMP) forms a complex with catabolite activator protein (CAP)
   c. The CAP-cAMP complex then binds to a portion of the *lac* operon promoter called the CAP site
   d. Binding at the CAP site enhances transcription initiation, resulting in increased enzyme production; thus, CAP-cAMP binding has a positive effect on transcription
   e. If glucose is present, the level of cAMP in the cell decreases, thereby reducing the amount of CAP-cAMP binding
3. The *lac* operon therefore is controlled in a positive and a negative fashion to ensure adequate enzyme production

## III. The *Trp* Operon: A Repressible System

### A. General information

1. *E. coli* cells require five enzymes to convert chorismic acid into the amino acid tryptophan
2. The genes that encode the tryptophan biosynthetic enzymes are grouped into a single transcriptional unit, called the tryptophan *(trp)* operon

3. If adequate levels of tryptophan are present in a cell, the enzymes are not needed and transcription is repressed
   a. A specific repressor protein, the *trp* repressor, can bind to the *trp* operon only when it forms a complex with tryptophan
   b. Tryptophan is called a *co-repressor* because it is needed for repressor binding
4. Because the operon is repressed only when the repressor protein is activated by a co-repressor, the operon is repressible

**B. Structure of the *trp* operon**

1. Like the *lac* operon, the *trp* operon contains linked structural genes and is adjacent to operator and promoter regions
   a. The structural gene *trpE* is located closest to the operator, followed by the genes *trpD, trpC, trpB,* and *trpA*
   b. The *trp* repressor is encoded by the independently transcribed regulatory gene, *trpR*
2. In addition, the *trp* operon contains a sequence between the operator and the *trpE* gene called *trpL* (the significance of this gene is discussed in part D. below)

**C. Co-repression**

1. When levels of tryptophan are insufficient, the *trp* repressor is not activated because the co-repressor is lacking, and the structural genes are transcribed
2. When levels of tryptophan are adequate, tryptophan serves as co-repressor and binds to the *trp* repressor at an allosteric site, resulting in a conformational change
3. The conformational change invoked by the binding of tryptophan enables the repressor to bind to the operator; thus, the repressor prevents RNA polymerase from initiating transcription
4. Thus, the enzymes required for the biosynthesis of tryptophan are present only when not enough tryptophan is available

**D. Attenuation in the *trp* operon**

1. Transcription of the *trp* operon also is regulated by a second mechanism called ***attenuation***
2. Attenuation depends upon the simultaneous transcription and translation of *trp* genes
3. The rate of ribosome movement during translation determines whether mRNA molecules are terminated prematurely or synthesized completely
   a. When transcription begins, the *trpL* region is the first to be transcribed; it yields a leader sequence (for an illustration, see *Attenuation in the* trp *Operon,* page 106)
      (1) The leader sequence has four subregions (1, 2, 3, and 4) that can form base pairs with one another in specific ways
      (2) Regions 2 and 3 are complementary; they pair and form a stem-loop structure
      (3) Regions 3 and 4 are complementary and form a stem-loop structure
   b. After RNA polymerase has completed transcription of the leader sequence and has moved on to the structural genes, ribosomes can initiate translation

## Attenuation in the *trp* Operon

Attenuation depends on the simultaneous transcription and translation of trp genes. In turn, the rate of translation determines whether messenger ribonucleic acid (mRNA) molecules are synthesized or terminated. If high levels of tryptophan are present, translation becomes rapid and transcription stops (as shown in top diagram). If low levels of tryptophan are present, translation is slow and transcription is completed (as shown in the bottom diagram).

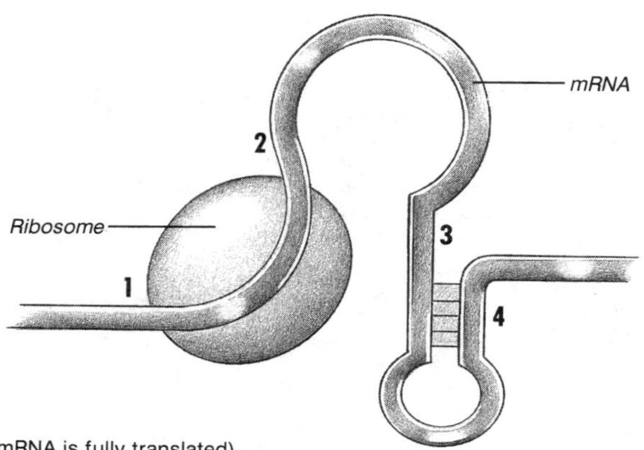


**1** (leader region of mRNA is fully translated)
**2** (segment of mRNA entering the ribosome)
**3** (segment of mRNA that forms base pairs with segment **4**)
**4** (segment of mRNA that forms base pairs with segment **3**)
Formation of hairpin structure (at segments **3** and **4**) results in termination of transcription

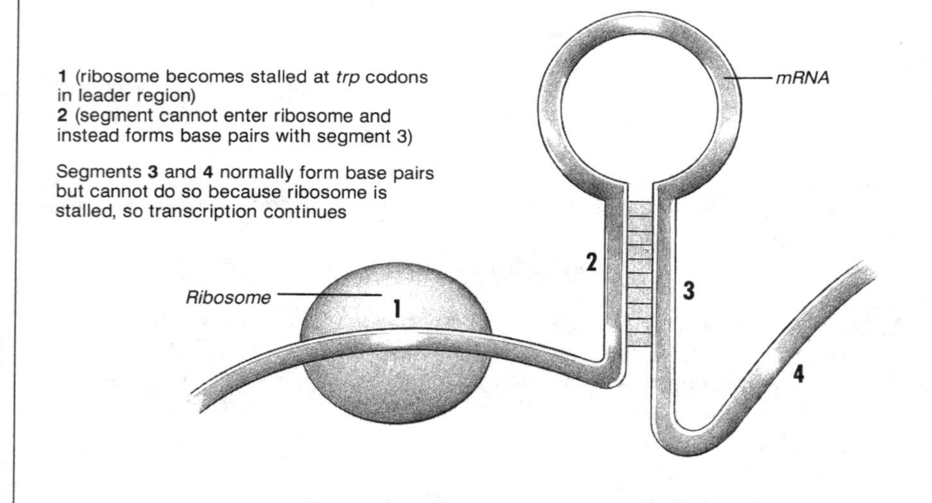


c. The leader sequence contains two tryptophan codons in region 1

d. If there is an adequate supply of tryptophan, *trp*-transfer RNAs (tRNAs) will be abundant and ribosome movement will be rapid

 (1) The ribosome rapidly moves through leader sequence region 1 and encounters region 2

(2) Because the ribosome is covering region 2, the only possible secondary structure of the leader sequence is the pairing of regions 3 and 4

(3) The stem-loop structure formed by such a pairing destabilizes the transcriptional apparatus and the RNA polymerase discontinues transcription

e. If tryptophan is in short supply, there will be few *trp*-tRNAs and the ribosome will slow down when it encounters the tryptophan codons

(1) The delay in ribosome movement will allow region 2 to pair with region 3 and form a stem-loop structure

(2) Because region 3 is paired with region 2, it cannot form a transcription termination stem-loop with region 4, and transcription continues

## IV. Lambda Bacteriophage: Multiple Operons

### A. General information

1. When lambda bacteriophage (phage) particles infect bacterial cells, the phage will enter the lytic cycle or the lysogenic cycle
2. In order for lytic development to occur, lambda phage genes must be expressed in a particular order (called a lytic cascade)
3. For lysogenic development to occur, lytic genes must be repressed

### B. Structure of the lambda phage chromosome

1. The genes for deoxyribonucleic acid (DNA) replication and lytic development are contained in the early right operon; genes required for lysogeny are located in the early left operon
2. Two promoters, $P_L$ and $P_R$, govern transcription of the left and right operons, respectively

a. Each of these promoters has three operators ($O_{L1}$, $O_{L2}$, $O_{L3}$ for $P_L$; $O_{R1}$, $O_{R2}$, $O_{R3}$ for $P_R$)

b. In addition to its effects on other genes, $P_L$ directs the leftward transcription of the *N* gene

(1) The N protein is called an anti-terminator protein

(2) When N is present, it binds to N utilization sites (*nut* sites) and facilitates the complete transcription of lytic genes

c. In addition to its effects on other genes, $P_R$ directs the rightward transcription of the *cro* (control of repressor and other), *cII*, and *Q* genes

3. Between the left and right early promoters lies the *cI* gene, which is controlled by its own promoters: $P_E$ establishes lysogeny and $P_M$ maintains lysogeny

a. The *cI* gene encodes the lambda phage repressor

b. Mutations in the *cI* gene prevent the establishment of lysogeny

### C. Lysis vs. lysogeny

1. When a cell is first infected by a lambda bacteriophage, transcription begins at $P_L$ and $P_R$, leading to the transcription of *N* and *cro* genes, respectively

2. Transcription from these promoters will continue and lead to the ultimate transcription of the lytic genes, unless repressor molecules bind to the left and right operators
   a. The transcription of a repressor is initiated by the binding of the *cII* gene product at $P_E$
   b. Once produced, a repressor will turn off transcription at $P_R$ and $P_L$
   c. In addition, a repressor stimulates its own production by binding at $P_M$
3. Therefore, a cell's decision to enter a lytic or lysogenic cycle involves a complex interaction between repressor and activator molecules

## Study Activities

1. Explain why operons containing genes required for anabolic reactions generally are repressible, whereas operons containing genes required for catabolic reactions generally are inducible.
2. Explain why operator mutations are *cis*-dominant and $I^s$ mutations are *trans*-dominant in the lac operon.
3. Describe the events that would occur at the *lac* operon if both lactose and glucose were present.
4. Distinguish between the effects of an inducer and a co-repressor on a repressor molecule.
5. Can attenuation be an effective gene control mechanism in eukaryotes? Explain why or why not.

# 13

# Control of Gene Expression in Eukaryotes

## Objectives

After studying this chapter, the reader should be able to:

- Discuss the possible levels of gene regulation in eukaryotes.
- Explain how the interaction of *cis*-acting elements and *trans*-acting factors influence gene expression in eukaryotes.
- Describe the genetic regulation of early development in *Drosophila*.
- Explain why *Caenorhabditis elegans* is a unique model system for the study of genetic development.
- Discuss how antibody diversity is achieved.

## I. Multiple Levels of Control

### A. General information

1. In Chapter 12, we considered various mechanisms of transcriptional control in prokaryotic cells
2. Recall that eukaryotic genes are not arranged in operons, but give rise to monocistronic messenger ribonucleic acid (mRNA) molecules
3. Although transcriptional control is important, additional levels of control add complexity to eukaryotic gene regulation
4. Eukaryotic control of gene expression consists of short- and long-term control
   a. Short-term control enables the cell to adjust rapidly to changes in the intra-cellular and intercellular environments
   b. Long-term control enables cells to differentiate, which allows for the development of organisms

### B. From DNA to protein

1. If we consider the steps involved in eukaryotic protein synthesis, it becomes obvious that many levels of control are possible
2. Transcriptional control includes regulation mechanisms that involve chromosome packaging, chemical modification of deoxyribonucleic acid (DNA), and placement of *cis*-acting regulatory regions
3. Posttranscriptional control includes mechanisms that involve mRNA processing and transport
4. Translational control includes mechanisms that involve mRNA-ribosome binding and mRNA degradation

5. Posttranslational control includes mechanisms that involve protein processing, transport, and degradation

## II. Transcriptional Control

### A. General information

1. As in prokaryotes, control of transcription in eukaryotes involves both *cis*-acting and *trans*-acting control mechanisms
2. Similarly, transcriptional control can be positive or negative

### B. Eukaryotic promoters

1. Nucleotide sequences at and near the TATA box and the CAAT box greatly influence the efficiency of transcription initiation; these sequences are called promoters
2. Promoter elements determine when transcription begins
   a. Different promoter elements determine which gene (or genes) will respond to specific promoter-binding molecules
   b. If binding of a molecule at a promoter element activates transcription, the element is a positive regulatory element; if binding of a molecule represses transcription, the element is a negative regulatory element

### C. Enhancers and silencers

1. Additional nucleotide sequences called ***enhancers,*** or upstream activating sequences, are located upstream from the promoter and are necessary for maximal rates of transcription
   a. Enhancers may be hundreds of base pairs (bps) away from the promoter
   b. Like promoters, enhancers bind regulatory proteins
2. Similar elements that repress transcription are called silencer elements, or ***silencers***
3. The net effect of regulatory sequences on transcription depends upon the regulatory proteins present in a particular cell
4. Transcription of different genes therefore is governed by specific combinations of regulatory elements

### D. *Trans*-acting factors

1. Remember that the rate of transcription depends on the presence or absence of *cis*-acting control regions and *trans*-acting proteins
2. Many eukaryotic transcription factors that bind to enhancer or promoter elements have been identified
   a. Some proteins, which specifically recognize CAAT and TATA boxes, must be present before RNA polymerase II can initiate transcription
   b. Transcription factors generally consist of two distinct domains, one that binds to DNA and one that initiates transcription
   c. DNA-binding domains are divided into subclasses, based on their structure
      (1) Proteins with the zinc finger motif form complexes with a zinc molecule and have projections that resemble fingers
      (2) Proteins with the helix-turn-helix motif contain two helices that fit into the grooves of a DNA molecule and are separated by a turn in the protein's secondary structure

(3) Leucine zipper proteins form dimers through hydrophobic interactions between leucine residues on both monomers; a DNA-binding region flanks each zipper region

(4) Helix-loop-helix proteins also form dimers, with the interacting regions containing two helices separated by a loop in the secondary structure (instead of leucine residues)

3. Steroid hormones also may serve as *trans*-acting proteins
   a. Only target cells that contain appropriate steroid hormone receptors can respond to a specific hormone
   b. Genes that are regulated by steroid hormones contain gene sequences to which the hormone-receptor complex binds
   c. In chicken oviducts, for example, estrogen binds to the enhancer of the ovalbumin (egg-white protein) gene with the aid of a receptor and activates transcription

### E. DNA methylation

1. After DNA replication, the addition of a methyl group chemically modifies a small percentage of bases
2. In some organisms (particularly mammals), transcriptionally inactive gene sequences contain a higher percentage of methylated cytosine (5-methylcytosine) residues than transcriptionally active gene sequences
3. Many studies suggest that cytosine methylation may be involved in the suppression of gene activity
4. Investigators caution, however, that these correlations do not reveal cause and effect relationships; additional evidence is needed to prove that methylation causes gene suppression

## III. Posttranscriptional Regulation

### A. General information

1. At the end of transcription, eukaryotic pre-mRNA molecules must be processed and transported into the cytoplasm for translation
2. After transport, the pool of mRNA molecules available for translation depends on the rate of degradation as well as the synthesis of additional molecules

### B. Differential mRNA processing and transport

1. Primary transcripts of some genes may be spliced in different ways (this is called ***alternative splicing),*** thereby creating more than one mature mRNA sequence
2. The site of polyadenylation (that is, the formation of the poly-A tail) may differ for transcripts produced in different tissues or at different times in an organism's development
3. Differential processing therefore can lead to different protein products arising from a single transcription unit
   a. For example, the same primary transcript gives rise to calcitonin in rat thyroid cells and calcitonin gene–related peptide in cells of the rat hypothalamus
   b. Differential processing also is important in the synthesis of immunoglobulin molecules (these molecules are described in section V. below)

4. The rate at which mRNA molecules are transported out of the nucleus may be influenced by the rate of splicesome assembly and dissociation
   a. RNA is retained in the nucleus as long as it is complexed with small nuclear ribonucleoprotein particles (snRNPs)
   b. mRNA molecules interact with the nuclear pores only after intron removal and splicesome dissociation

**C. mRNA stability**

1. Structural differences in individual mRNA molecules cause differences in mRNA stability
2. mRNA degradation depends upon the interaction of nucleases and the mRNA structure
3. For example, regions rich in adenine (A) and uracil (U) in 3′ trailing sequences are linked to mRNA instability

## IV. Gene Regulation in Development and Differentiation

**A. General information**

1. Because the DNA content of an organism's cells remains constant from cell to cell and throughout development, cellular differentiation and organismal development must be the result of a programmed sequence of gene activation and repression
2. Studies of the fruit fly, *Drosophila melanogaster,* and the nematode, *Caenorhabditis elegans,* have elucidated various aspects of developmental genetics
3. Gene interactions observed in both wild-type and mutant examples of these two species have greatly increased our understanding of how animal body plans are established

**B. Genetic control of *Drosophila* development**

1. Because of its short-lived generations, small genome, low chromosome number (n = 4), and polytene chromosomes, *Drosophila* is an excellent choice for studying the genetic control of early animal development
   a. A fertilized egg becomes a larva, which undergoes three molts and becomes a pupa; the pupa, in turn, becomes an adult fly
   b. The entire process — from egg to adult — takes approximately 9 days
2. Knowledge about the early developmental events in *Drosophila* is important to the understanding of genetic control mechanisms
   a. Before fertilization, gradients of molecules are developed within *Drosophila* eggs
      (1) The posterior pole contains the polar cytoplasm, in which small polar granules of RNA and protein can be seen
      (2) The female pronucleus is centrally located in the egg
   b. After fertilization, the zygote rapidly divides; 9 cycles of division ensue in the absence of cytokinesis
      (1) The multinucleated structure is called a ***syncytium***
      (2) After 9 divisions, nuclei migrate to the periphery of the egg
         (a) Some nuclei migrate into the polar cytoplasm, where they are surrounded by invaginations in the plasma membrane; these are called *pole cells*

(b) The pole cells contain polar gametes and represent the precursors of the ***germ line,*** cells that will ultimately give rise to gametes

(c) The remainder of the nuclei undergo additional divisions at the egg periphery and form the syncytial blastoderm

c. During the blastoderm stage, membranes develop around the peripheral nuclei, leading to the formation of somatic cells

(1) Two types of cells are present in the blastoderm: one that will produce larval tissues and one that will eventually produce adult tissues

(2) Cells that are destined to become adult tissues (such as genitalia, wings, and eyes) remain in groups of undifferentiated cells while the larva matures

(3) A group of undifferentiated cells is called an ***imaginal disc;*** each disc develops into a specific adult organ

d. Gastrulation occurs as the ventral midline invaginates; during this process, pole cells migrate to a position consistent with that of the gonads

e. Body segments soon become apparent and organogenesis occurs

(1) Development of body structures depends upon two molecular gradients, one along the posterior-anterior axis and another along the dorsal-ventral axis

(2) The position of cells within these gradients determines the cell's developmental fate

3. Genes involved in developmental control are defined by mutations that result in the formation of abnormal structures or death; these genes fall into one of three major classes

a. Maternal genes (those expressed by the mother) are responsible for protein gradients that develop in the egg and therefore are involved in spatial organization of the early embryo

(1) mRNA from maternal genes is synthesized in cells that surround the egg; then, it is transported into the egg

(2) mRNA from maternal genes is retained in the anterior or posterior region of the egg

(3) The *bicoid* gene is required for normal development of anterior segments

(4) The *nanos* gene is responsible for normal development of posterior segments

b. Segmentation genes are transcribed in the zygote and are responsible for the number and organization of segments

(1) Approximately 20 segmentation genes exist; these are further subdivided into gap genes, pair-rule genes, and segment-polarity genes

(2) Segmentation genes are involved in a cascade of gene expression: the expression of gap genes (which defines four embryonic regions) is followed by expression of pair-rule genes that define segment pairs; this is followed by expression of segment-polarity genes, which define individual segments

c. Homeotic genes, which are expressed after segmentation genes, determine the identity of individual segments

(1) Mutations in these genes, called ***homeotic mutations,*** result in the development of normal structures at inappropriate segments

(2) Two major homeotic gene complexes (gene groups) have been identified: the *Antennapedia* complex and the *bithorax* complex

(3) Genes in the *Antennapedia* complex control the identity of anterior segments, while genes in the *bithorax* complex control the identity of posterior segments

(4) Mutation in the *Antennapedia* complex can result in the development of legs in place of antennae, for example; mutation in the *bithorax* complex can result in development of extra wing pairs

(5) All homeotic genes share a 180-bp DNA sequence called the homeobox, which gives rise to a DNA-binding site (with a helix-turn-helix motif) on all proteins encoded by such genes

(6) Homeoboxes have been discovered in many eukaryotes, including humans, and may represent a key to the understanding of developmental regulation

### C. The development of *Caenorhabditis elegans*

1. Because of its small size (about 1,000 cells), *C. elegans* is a unique organism in which to study cell fate
2. By observing all stages of development, biologists have been able to determine the precise lineage of every cell in the adult nematode, thereby facilitating the construction of a complete map of the cells' fates
3. By selectively killing specific cells, one can study the regulation of cell position and cell fate
4. Patterns of cell lines have been found to be *invariant,* meaning that surrounding cells cannot provide the missing function if a cell is killed
5. Knowledge of normal cell lineage patterns also is extremely useful in deciphering the effects of mutation

## V. Immunogenetics

### A. General information

1. The immune system provides protection against infectious agents in vertebrates
2. The immune system recognizes foreign molecules; molecules that elicit a response from the immune system are called *antigens*
3. Many types of cells are involved in the immune response
   a. B cells are white blood cells that secrete ***antibodies,*** which are protein molecules that bind to antigens
   b. Another cell, the cytotoxic T lymphocyte, recognizes host cells that have been infected
4. The immune system's ability to recognize and respond to the enormous number of antigens encountered by an individual in a lifetime is due in large part to the diversity of antibodies, or ***immunoglobulins,*** produced by the body

### B. Immunoglobulins

1. Different immunoglobulins are produced by B cells, each with slightly different properties; they are IgG, IgA, IgM, IgD, and IgE
2. All immunoglobulin molecules are proteins consisting of two light (L) polypeptide chains and two heavy (H) polypeptide chains that are joined together by disulfide bonds
   a. Five primary types of H chains exist, one for each type of immunoglobulin: $\gamma$ (gamma), $\alpha$ (alpha), $\mu$ (mu), $\delta$ (delta), and $\varepsilon$ (epsilon)

## Rearrangement and Expression of Gene Segments Encoding the κ Light Chain in Mice

This rearrangement represents one of many possible recombinations that produces antibody diversity.

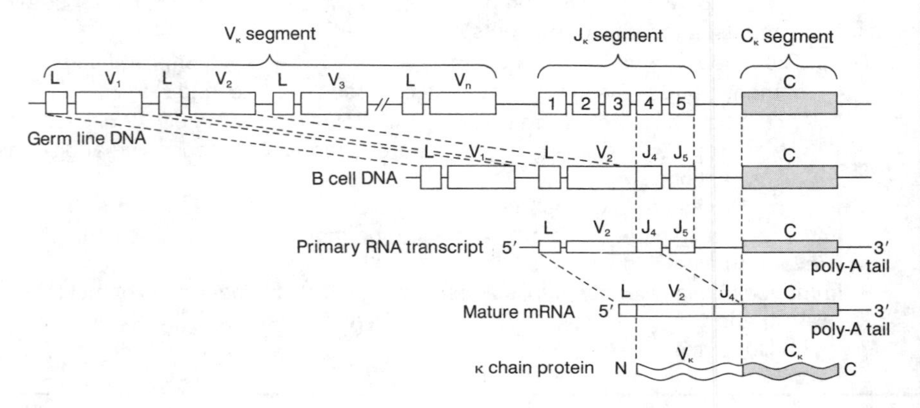


From *Genetics* (3rd ed.) by Peter Russell. ©1992 by Peter Russell. Reprinted with permission of HarperCollins.

b. Two types of L chains exist: λ (lambda) and κ (kappa)

3. Each polypeptide chain, light and heavy, contains a region of variable amino acid sequence (V) and a region of constant amino acid sequence (C)

**C. Generation of antibody diversity**

1. Variability in light and heavy chains is the result of variability in DNA level that is generated by somatic recombination
2. The assembly of the κ light chain in the mouse, for example, involves three types of DNA segments
   a. The L-$V_\kappa$ segment encodes a leader sequence (which is required for immunoglobulin secretion) and most of the V region of the mature light chain; mice have approximately 350 L-$V_\kappa$ segments
   b. The $C_\kappa$ segment encodes the C region of the mature light chain; mice have one $C_\kappa$ segment
   c. The *J segment,* or joining segment, links segments together; the $J_\kappa$ joins the $V_\kappa$ and $C_\kappa$ segments (mice have four $J_\kappa$ segments)
3. In immature B cells, all L-$V_\kappa$ segments, $J_\kappa$ segments, and the $C_\kappa$ segment are arranged in tandem on the mouse chromosome (see *Rearrangement and Expression of Gene Segments Encoding the* κ *Light Chain in Mice*)
4. During maturation of the B cell, one L-$V_\kappa$ segment becomes attached to one $J_\kappa$ segment and the $C_\kappa$ segment via somatic recombination
   a. In V-J joining, a single crossover in a stem-loop structure removes intervening L-V and J segments
   b. The site of V-J joining is variable, resulting in the creation of additional immunoglobulin diversity; this is called junctional diversity

5. After transcription, the region separating the selected $J_\kappa$ segment and $C_\kappa$ segment is removed and the mRNA is translated
6. Similar recombination takes place during the synthesis of λ light chains and heavy chains
   a. The mouse λ light chain complex contains two L-$V_\lambda$ segments, four $J_\lambda$ segments, and four $C_\kappa$ segments
   b. The λ heavy chain complex contains about 200 L-V segments, 4 J segments, and 8 C segments of 5 major types (γ, α, μ, δ, and ε)
      (1) The λ heavy chain complex also contains approximately 12 diversity (D) segments; diversity segments are responsible for encoding additional antibody variability
      (2) D-J joining precedes V-DJ joining
7. Immunoglobulin gene complexes appear to be more susceptible to mutation *(hypermutable)* in specific regions
8. Immunoglobulin diversity thus is a result of variability in sequences of multiple L-V, J, and C segments; junctional diversity; and hypermutability of immunoglobulin genes

---

## Study Activities

1. Which mechanisms of gene regulation in eukaryotes are similar to those found in prokaryotes? Which are different?
2. Compare and contrast the following *cis*-acting structures: positive regulatory promoter element, negative regulatory promoter element, enhancer, and silencer.
3. What is alternative splicing and how does it influence protein synthesis?
4. Describe the hierarchy of gene expression that controls development in *Drosophila* embryos.
5. Homeotic mutations also have been discovered in plants. Applying what you know about the nature of these mutations, what phenotypes might be displayed by homeotic flower mutations?
6. Describe the various mechanisms that ensure that an individual will have a diversity of antibodies.

# 14

# Gene Mutation

## Objectives

After studying this chapter, the reader should be able to:
- Classify mutant phenotypes.
- Explain the molecular changes that are responsible for gene mutation.
- Describe various mechanisms of spontaneous gene mutation.
- Discuss the various mechanisms by which specific mutagens induce gene mutation.

## I. The Nature of Mutation

### A. General information

1. Geneticists rely heavily on the occurrence of mutations for genetic analyses
2. Although the genetic studies described in earlier chapters provided information about genetic mutations, none of the studies was specifically designed to elucidate the molecular basis of mutation or the mechanisms of genetic change
3. Two types of mutation exist: gene mutation and chromosome mutation
   a. Gene mutation involves any one of a number of changes in a deoxyribonucleic acid (DNA) sequence at a specific locus
   b. Chromosome mutation, which is discussed in Chapter 15, involves a change in the organization of a chromosome or chromosomes
4. Any mutation away from a wild-type allele is considered a *forward mutation;* a mutation back to the wild-type allele is called a *reverse mutation*

### B. Spontaneous vs. induced mutation

1. Mutations that occur spontaneously at a very low rate are *spontaneous mutations*
   a. Spontaneous mutation is caused by several factors, such as errors in DNA replication and spontaneous DNA lesions
   b. Most human hereditary diseases, such as cystic fibrosis and sickle cell anemia, result from spontaneous mutation
2. Mutations also are induced by a variety of agents; these are called *induced mutations*
3. Induced mutation involves DNA base replacement, base alteration with subsequent mispairing, or loss of base pairs with subsequent blockage of replication
   a. Hermann Muller, who discovered that X-rays cause mutations in 1927, is credited with first recognizing that mutations can be induced
   b. DNA sequencing reveals that each agent that induces mutation, called a mutagenic agent or ***mutagen,*** produces its own spectrum of mutations

## C. Somatic vs. germinal mutation

1. Mutations can occur in any cell, including somatic cell and germ cells
2. If a cell in which a somatic mutation has occurred undergoes mitotic division, a population of identical, mutant cells will arise; this population of cells is called a ***clone***
   a. A clone commonly is visible as a *mutant sector,* a group of mutant cells on a wild-type background
   b. The size of a mutant sector depends upon the timing of the event that causes the mutation
      (1) If the mutation occurs early in development, the mutant sector will be large
      (2) If the mutation occurs late in development, the mutant sector will encompass relatively few cells
   c. Somatic mutations are not passed on to offspring; however, in organisms that do not have a distinct germ line (such as plants), a mutant somatic cell may eventually give rise to mutant germinal tissue, which could lead to sexual transmission of the mutation
3. Germinal mutation may lead to mutant gametes and, subsequently, mutant zygotes
   a. Individuals with germinal mutations may have a wild-type phenotype
   b. If a germinal mutation is recessive, it may not manifest itself for many generations

## D. Random vs. directed mutation

1. Geneticists have wondered whether mutation is a random event or is directed by environmental cues
2. Two classic experiments in bacterial genetics support the random nature of mutation
   a. In 1943, Salvador Luria and Max Delbrück compared the frequency with which bacterial cells resistant to bacteriophage T1 occurred during a culture
      (1) Twenty individual liquid cultures of *Escherichia coli* were transferred onto petri dishes containing a dense layer of T1 phage particles
      (2) Variation in the number of resistant colonies per culture (each presumably arising from a single, resistant bacterium) led the researchers to conclude that resistance was the result of random mutation rather than a change directed by the presence of T1 phages in the culture environment
   b. In 1952, Joshua Lederberg and Esther Lederberg demonstrated that mutant bacterial cells can arise prior to growth in a selective environment
      (1) Cells were grown on petri dishes in a nonselective medium that contained all necessary nutrients but no antibiotics or phage particles
      (2) Cells were then transferred by ***replica plating*** to plates that contained a selective medium
         (a) Replica plating is performed by touching the surface of a plate containing bacterial cells with a sterile velvet cloth and then touching the cloth to a second, sterile plate
         (b) Replica plating transfers colonies from one plate to another while maintaining their relative positions

(3) The Lederbergs found that mutant colonies that grew on the selective medium always occurred at the same position on multiple replica plates, leading them to conclude that the mutations arose before growth in the selective environment

3. Mutation events therefore are random and are not directed by specific environmental conditions

**E. Mutation rates and frequency**

1. The measurement of a gene's tendency to mutate is called the *mutation rate*
   a. In single-celled organisms, the mutation rate equals the number of mutant cells that arise per cell division; in multicelled organisms, the mutation rate equals the number of mutations that arise per generation
   b. Spontaneous mutation rates are highly variable; in humans, the rate for individual genes generally is between $1 \times 10^4$ and $1 \times 10^6$ per generation, depending upon the locus involved
2. *Mutation frequency* is the frequency at which a specific mutation is found in a population of cells, gametes, or individuals

## II. Mutant Phenotypes

**A. General information**

1. Mutant phenotypes are classified in a number of ways; for example, they can be categorized according to the visible manifestation of the mutant allele or the way the mutation affects function
2. Any classification system, however, is not precise, and many mutations can be placed in two or more categories

**B. Morphological mutations**

1. Morphological mutations visibly change the appearance of an organism
2. Changes in an organism's color, body size and shape, and bacterial colony size are examples of morphological mutations
3. Albinism can be considered a morphological mutation

**C. Biochemical mutations**

1. Biochemical mutations result in an alteration in or loss of biochemical function
2. Bacterial auxotrophs that require nutritional supplementation for survival are examples of biochemical mutations
3. Phenylketonuria, the inability to convert phenylalanine into tyrosine, is an example of a biochemical mutation that occurs in humans
4. Some biochemical changes result in resistance to a specific antibiotic or other growth inhibitor and are aptly termed resistance mutations

**D. Conditional mutations**

1. Some mutant phenotypes are expressed only in certain environments; the mutations that produce these phenotypes are called conditional mutations
   a. The condition under which a wild-type phenotype is expressed is called the permissive condition
   b. The condition under which a mutant phenotype is expressed is called the restrictive condition

2. Temperature-sensitive mutant organisms, for example, express the mutant phenotype only when grown at higher temperatures

**E. Lethal mutations**

1. A lethal mutation reduces an organism's ability to survive
2. Lethal alleles (described in Chapter 3, Advanced Mendelian Analysis) may exert their effects at various times during development
3. By their very nature, lethal mutations are difficult to isolate
4. Conditional lethal mutations simplify the maintenance of mutant stocks in laboratories, and lend themselves to the study of mutant effects when the affected organism is transferred to restrictive conditions

**F. Hypermorphic, hypomorphic, and amorphic mutations**

1. Hypermorphic, hypomorphic, and amorphic mutations depend on comparison of the mutant phenotype with the wild-type (standard) phenotype
2. Hypermorphs display a phenotype that is "more than" the phenotype exhibited by the wild type; a mutant bacterial strain, for example, may overproduce a specific metabolite
3. Hypomorphs display a phenotype that is "less than" the phenotype exhibited by the wild type; humans with familial hypercholesterolemia, for example, have half of the normal number of cellular receptors required for normal cholesterol metabolism
4. Amorphs lack a specific wild-type function or attribute; *Drosophila* that express the white-eye mutation, for example, have no eye pigmentation

## III. Molecular Basis of Gene Mutation

**A. General information**

1. DNA changes underlying a mutant phenotype may involve thousands of nucleotide pairs or a single pair of nucleotides
2. A single base pair substitution is called a *point mutation*
3. Although point mutations alter the sequence of nucleotides, they may not result in a mutant phenotype
   a. Because of the degeneracy of the genetic code, nucleotide substitutions may not result in an alteration in the amino acid sequence; such a mutation is called a *silent mutation*
   b. Although a point mutation may result in a change at the protein level, the new codon may specify an amino acid that is chemically equivalent to the original so that a *neutral mutation* — which does not functionally alter the resultant protein — occurs

**B. Transitions and transversions**

1. Base substitutions in which one purine is substituted for another purine or one pyrimidine is substituted for another pyrimidine are called ***transition mutations***, or transitions
2. Four possible transitions exist: AT → GC, GC → AT, TA → CG, and CG → TA
3. Base substitutions in which a purine is substituted for a pyrimidine or vice versa are called ***transversion mutations***, or transversions
4. Examples of transversions include: AT → TA, AT → CG, and CG → GC

5. A base pair change that alters the messenger ribonucleic acid (mRNA) codon such that a different amino acid is inserted in the protein is called a ***missense mutation***
6. A base pair change that alters the mRNA codon such that a termination codon is formed is called a ***nonsense mutation***
   a. A change in the DNA coding strand from 3′-GTT-5′ to 3′-ATT-5′ would change the mRNA codon from 5′-CAA-3′ (glutamine) to 5′-UAA-3′ (terminator)
   b. A nonsense mutation prematurely terminates translation and commonly gives rise to a nonfunctional protein

**C. Frameshift mutations**

1. If a nucleotide pair is added or deleted from a gene sequence, the sequence of codons in the resultant mRNA will be altered; this type of mutation is called a ***frameshift mutation***
2. For example, if a frameshift mutation results in a change in the mRNA sequence from 5′-AUUCGCCGACAG-3′ to 5′-AAUUCGCCGACAG-3′, the protein sequence will change from isoleucine-arginine-arginine-glutamine to asparagine-serine-proline-threonine
3. Frameshift mutations drastically alter protein sequences because they theoretically can alter every codon that follows the site of mutation

**D. Duplications and deletions**

1. Duplications and deletions, as the names imply, involve the addition or loss of many base pairs
2. These mutations generally arise in regions of repeated DNA sequences, where replication and recombination errors may occur
3. Deletions and duplications may lead to frameshift mutations
4. Large-scale duplications and deletions (at the chromosome level) will be considered in Chapter 15, Chromosome Mutation

**E. Reverse mutations and suppressor mutations**

1. A *reverse mutation* can restore a mutant protein to full, or nearly full, function
2. Two types of reversion events can occur: exact and equivalent
   a. An exact reversion mutation restores the mutant gene to its original DNA sequence: UUA (leucine) → AUA (isoleucine) → UUA (leucine)
   b. An equivalent reversion mutation restores the protein to its original (or equivalent) sequence, but does not restore the exact DNA sequence: UUA (leucine) → AUA (isoleucine) → CUA (leucine)
3. The effects of a mutation may be partially or fully eliminated by the occurrence of a second mutation, called a ***suppressor mutation***
4. Two primary types of suppressor mutations exist: intragenic and extragenic (also called intergenic)
   a. Intragenic suppressor mutations occur in the same gene in which the original mutation occurred
      (1) A point mutation can be suppressed if a second mutation alters another nucleotide of the codon in which the original mutation occurred, resulting in the wild-type (or equivalent) amino acid: UUA (leucine) → UUU (phenylalanine) → CUU (leucine)

(2) A frameshift mutation that results from deletion of a nucleotide pair may be suppressed if a second mutation adds a nucleotide pair in the vicinity of the original mutation

b. Extragenic suppressor mutations occur in a different gene than the one in which the original mutation occurred

(1) A nonsense mutation can be suppressed by the mutation of transfer RNA in which the anticodon pairs with the termination codon and inserts an amino acid, thus allowing the mRNA to be fully translated

(2) A physiological suppressor mutation circumvents a defect in one biochemical pathway by altering a second biochemical pathway

## IV. Mechanisms of Gene Mutation

### A. General information

1. Mutations can result from a variety of spontaneous or induced processes
2. Although spontaneous mechanisms were once thought to be the result of indigenous mutagens (such as radiation), evidence has disproved this theory

### B. Replication errors

1. DNA bases can occur in a variety of forms called tautomers; a tautomer is an isomer whose atoms and bonds occupy different positions)
   a. Although the "normal" form of DNA bases is the keto form, these bases also exist in the imino and enol forms in rare instances
   b. The imino and enol forms do not consist of the same nucleotide pairs as the keto form
      (1) The imino form of cytosine can pair with adenine, for example
      (2) The enol form of guanine can pair with thymine
2. Mispairing can lead to inappropriate nucleotide incorporation by DNA polymerase
   a. Tautomeric shifts produce transition but not transversion mutations
   b. Transition mutations also can occur when ionized bases are present
3. Errors in replication also may lead to frameshift mutations
   a. In regions of repeated DNA, the nascent DNA strand can slip one or more nucleotides to the left or right, where it is stabilized by base-pairing
   b. Completion of replication results in the addition or deletion of nucleotide pairs
   c. Unless the number of altered nucleotide pairs is a multiple of 3, a frameshift mutation results

### C. Spontaneous lesions

1. Spontaneous DNA damage may be the result of several mechanisms, including depurination, deamination, and oxidative damage
2. Depurination occurs when the glycosidic bond is disrupted between deoxyribose and the attached base
   a. Apurinic sites (those that do not contain a purine) cannot specify base incorporation during replication and may lead to base substitution
   b. Repair mechanisms preferentially insert adenine across from apurinic sites, thus leading to GC → TA and AT → TA transversions
3. Deamination changes one base into another, causing a point mutation
   a. Deamination of 5-methylcytosine yields thymine
   b. Deamination of cytosine yields uracil

4. Oxidative DNA damage is caused by various metabolic products, including hydrogen peroxide, hydroxyl radicals, and superoxide radicals; the products of oxidative damage can cause base substitution to occur

**D. Incorporation of base analogs**

1. Some mutagens exert their effects after they are incorporated into replicating DNA
2. Chemicals that closely resemble normal DNA bases and are incorporated by DNA polymerase are called ***base analogs***
3. Once incorporated into replicating DNA, base analogs form inappropriate base pairs
   a. In the enol and ionized forms of DNA, 5-bromouracil, a thymine analog, forms a base pair with guanine
   b. In the keto form of DNA, the adenine analog 2-aminopurine can pair with thymine or cytosine
   c. Both 5-bromouracil and 2-aminopurine cause transition mutations

**E. Chemically induced base modification**

1. Some mutagens cause base alterations that lead to mispairing and base substitution
2. Base-modifying agents may act as deaminating agents, hydroxylating agents, or alkylating agents
   a. Nitrous oxide removes amino groups from guanine, cytosine, and adenine
   b. When hydroxylamine adds a hydroxyl group to cytosine, the cytosine pairs with adenine rather than guanine
   c. Methylmethane sulfonate and ethylmethane sulfonate add alkyl groups to bases

**F. Intercalating agents**

1. Because their molecules are planar, some chemical mutagens can intercalate (insert themselves) between adjacent bases
2. Once inserted, intercalating agents disrupt normal DNA replication and lead to nucleotide insertions and deletions
3. Proflavine, acridine orange, ethidium bromide, and dioxin are intercalating agents

**G. Radiation**

1. Nonionizing radiation (such as ultraviolet light) and ionizing radiation (such as X-rays) can cause mutations
2. Ultraviolet light induces photochemical modification of DNA, primarily the formation of thymine dimers
   a. Thymine dimers form between adjacent nucleotides and interrupt normal base-pairing
   b. Repair mechanisms (which are discussed in Chapter 16, Mechanisms of DNA Repair and Recombination) frequently insert incorrect bases across from ultraviolet photoproducts, thereby introducing transitions and transversions
3. Ionizing radiation can penetrate tissues, releasing electrons as it travels through cells
   a. Ionized bases may not pair correctly, thereby leading to gene mutation
   b. Ionizing radiation also can lead to chromosome breakage

c. Exposure to radiation can result in a higher frequency of stillbirths, major congenital defects, malignancies, and chromosomal aberrations

## Study Activities

1. Define and give examples of the following: forward mutation, reverse mutation, mutagen, mutation rate, and mutation frequency.
2. Cite several examples of each of the following types of mutation: morphological, biochemical, conditional, hypermorphic, hypomorphic, and amorphic.
3. Which of the following DNA changes is most likely lead to a nonfunctional protein: a transition mutation that changes a CCU codon into a CCC codon; a transversion mutation that results in a silent mutation; a spontaneous frameshift mutation; or an induced neutral mutation. Explain your answer.
4. Distinguish between exact reversion and equivalent reversion.
5. Make a table that summarizes the mechanisms of spontaneous and induced mutation; also include appropriate mutagens.

# 15

# Chromosome Mutation

## Objectives

After studying this chapter, the reader should be able to:

- Describe the various aspects of eukaryotic chromosome topography that are useful in cytological detection of chromosome mutation.
- Prepare chromosome diagrams that illustrate the concepts of deletion, duplication, inversion, and translocation.
- Draw chromosome pairing relationships observed in individuals who are heterozygous for structural chromosome changes.
- Compare and contrast the genetic ramifications of deletions, duplications, inversions, and translocations.
- Distinguish between the concepts of haploid chromosome number *(n)* and monoploid chromosome number *(x)*.
- Utilize appropriate genetic terminology when discussing changes in chromosome number.
- Identify mechanisms that may lead to changes in chromosome number.
- Cite examples of chromosome mutation in humans.

## I. Chromosome Morphology

### A. General information

1. Gross changes in eukaryotic chromosome structure and number are cytologically detected and characterized with the aid of distinct chromosome landmarks
2. Aspects of chromosome morphology that are useful to cytogeneticists include chromosome size, location of constrictions, and banding patterns

### B. Chromosome size

1. Chromosome size may or may not differ from one chromosome to another in a genome
2. In human cells, the longest chromosome (#1) is approximately four times the size of the smallest (#21)
3. However, chromosomes rarely are distinguished solely on the basis of size

### C. Primary constriction

1. Because of their physical appearance, centromeres are called the primary constriction

2. Centromeres divide each chromosome into two "arms" and are the site of microtubule attachment during nuclear division
   a. A chromosome whose centromere is in the middle and has two arms of equal length is called a ***metacentric chromosome***
   b. A chromosome whose arms are of slightly different lengths is called a ***submetacentric chromosome***
   c. A chromosome with a centromere that is close to one end of the chromosome is called an ***acrocentric chromosome***
   d. A chromosome with a centromere that is at the end of the chromosome is called a ***telocentric chromosome***

**D. Secondary constrictions**

1. Genes that encode ribosomal ribonucleic acid (rRNA) are arranged in tandem repeats at one or more loci
2. After transcription, rRNA forms an organized structure called the nucleolus that remains attached to the chromosome at the site of the rRNA genes
3. The attachment site of the nucleolus is called the ***nucleolus organizer region (NOR);*** because of its constricted appearance, this site also is called the secondary constriction
4. Depending upon the species, a haploid genome may have one or many secondary constrictions
5. Like primary constrictions, secondary constrictions serve as useful cytological landmarks

**E. Banding patterns**

1. Specialized staining techniques (introduced in Chapter 6, Structure of the Eukaryotic Chromosome) are useful in visualizing differences in the chemical composition and genetic state of chromatin
2. G-banding and Q-banding are valuable techniques used in medical genetics because banding patterns are unique for each human chromosome
   a. Metaphase chromosomes can be stained, photographed, individually cut out of a photographic print, and arranged in pairs by size and banding pattern
   b. A chromosome complement that is arranged in this way is called a ***karyotype***
   c. Specific chromosomal areas are designated by the chromosome number, the position on the chromosome (short or long arm, designated p and q, respectively), the region on the arm, and the band within the region
   d. For example, the beta hemoglobin gene is located on the short arm of chromosome 11 in region 15 at band 5; this is written as 11p15.5

## II. Changes in Chromosome Structure

**A. General information**

1. Chromosome mutation that involves breakage of the chromosome can result in the deletion, duplication, or rearrangement of large chromosome segments
2. Light microscopy generally can detect these rearrangements (see *Chromosomal Changes*)
3. Observation of aberrant chromosome structure and behavior can yield valuable insights into the normal structure and behavior of chromosomes

## Chromosomal Changes

Diagrammatic representations of duplicated, deleted, inverted, and translocated chromosomes are shown below.

**Duplication**
The *GH* segment of the chromosome is duplicated.

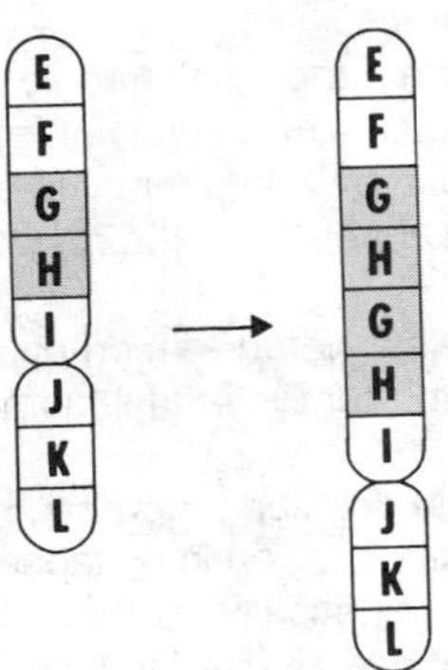

**Deletion**
The *G* segment of the chromosome is deleted.

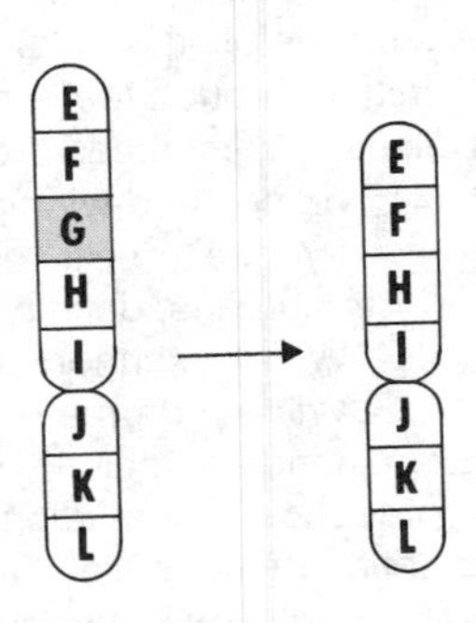

**Pericentric inversion**
The *EFG* segment, along with the chromosome's centromere, is inverted.

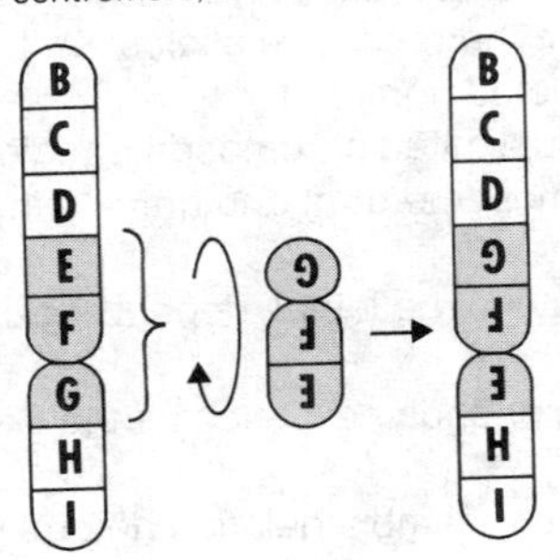

**Paracentric inversion**
The *BCD* segment is inverted; the centromere is not included.

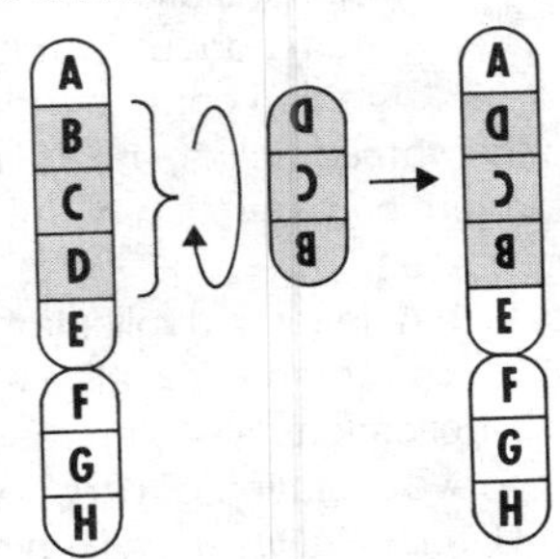

**Reciprocal translocation**
Two chromosomes interchange segments.

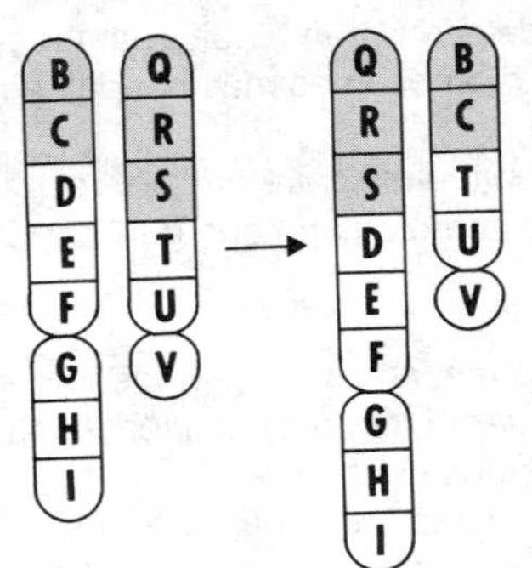

4. This cytological approach to genetics is called cytogenetics

**B. Deletions and duplications**

1. Two types of ***deletions,*** or deficiencies, exist: terminal deletions, which are produced by a single chromosome break, and interstitial deletions, which are produced by two chromosome breaks
   a. Deletions, which typically cause abortion, are expressed as gametic abortion in plants and zygotic abortion in animals
   b. A small deletion or a larger deletion for a nonessential region may be transmitted through the female parent in plants or a surviving zygote in animals
   c. Deletions sometimes are associated with unique phenotypes
      (1) The notched-wing phenotype of *Drosophila* is the result of heterozygosity for a specific deletion
      (2) In humans, deletions have been associated with cri-du-chat syndrome, Wolf-Hirschhorn syndrome, Wilms' tumor, retinoblastoma, and Prader-Willi syndrome
   d. Examination of meiotic chromosomes in deletion heterozygotes reveals the presence of a deletion loop, an unpaired region in synapsed homologs
   e. Deletion mutants also can be detected genetically
      (1) A deletion heterozygote would not be expected to produce recombinants in the deficient region
      (2) Deletion mutations do not revert back to wild-type structures
      (3) Deletions allow for the phenotypic expression of recessive alleles on the homologous chromosome when the recessive alleles are located in a region corresponding to the deleted homolog
2. As the name implies, a ***duplication*** is a duplicated chromosome region in a tandem configuration on the same chromosome or at another site in the genome
   a. Like deletions, duplications may or may not be lethal, depending upon the size of the duplicated region
   b. Duplications also can result in specific phenotypes, such as the Bar (slit-like) eye mutation of *Drosophila*
   c. Cytologically, duplication heterozygotes resemble deletion heterozygotes in that the paired homologs exhibit an unpaired loop
   d. Duplication of gene sequences plays an important role in the evolution of gene families
3. Duplications and deletions can result from the process of unequal crossing-over
   a. Homologous chromosomes may not pair accurately in regions of repeated DNA sequences
   b. Crossing-over in the mispaired region results in the production of one deletion-carrying homolog and one duplication-carrying homolog

**C. Inversions**

1. An ***inversion*** is the reversal of a chromosome segment and, subsequently, the reversal of the order of genes contained in that segment
2. Two types of inversions exist
   a. Rearrangements that are confined to a single arm of a chromosome and do not include the centromere are called ***paracentric inversions***
   b. Rearrangements that span both arms and therefore include the centromere are ***pericentric inversions***

3. Inversions can exist in both homozygous or heterozygous conditions
   a. Homozygous inversions behave like normal chromosomes, except that new linkages are established
   b. Although heterozygous inversions are viable, they result in sterility — the extent of which depends on the size of the inverted segment
4. Meiotic pairing of heterozygous inversions depends upon the size of the inverted segment
   a. If the inversion is short, the segment may remain unpaired; also, nonhomologous segments sometimes pair
   b. If the inversion is long, the chromosomes synapse in such a manner that homologous loci are paired and form a looped structure called an inversion loop
5. Crossing-over within the inversion loop of both pericentric and paracentric inversions results in the formation of duplication-carrying and deletion-carrying chromosomes
   a. Because the products of recombination are not consistent with viability, crossover products are not recovered
   b. The recombination frequency for genes within an inverted region therefore is zero
6. Crossing-over within paracentric inversions also involves the formation of dicentric bridges and acentric fragments
   a. One of the homologs produced from crossing-over within a paracentric inversion is *dicentric* (that is, having two centromeres); the two centromeres will stretch the chromosome from pole to pole when they separate during anaphase I
   b. The other homolog is *acentric* (without a centromere) and is unable to migrate to the other pole

**D. Translocations**

1. The breakage of two nonhomologous chromosomes, followed by mutual exchange of chromatin, is called a reciprocal ***translocation*** (or interchange)
2. Like inversions, translocations exist in heterozygous or homozygous conditions
   a. Homozygous translocations behave like normal chromosomes, except that new linkages are established and chromosome sizes may be altered
   b. Although heterozygous translocations are viable, they result in varying degrees of sterility and the establishment of new linkages
3. Heterozygous translocations are recognizable cytologically by the cross-shaped pachytene configurations that involve four chromosomes: normal chromosome 1 (N1), normal chromosome 2 (N2), translocated chromosome 1 (T1), and translocated chromosome 2 (T2)
4. Disjunction of chromosomes at anaphase I in a translocation heterozygote can occur via one of three methods: adjacent-1 segregation, adjacent-2 segregation, and alternate segregation
   a. Adjacent-1 segregation involves the movement of adjacent, nonhomologous centromeres to the same pole (N1 and T2 to one pole, N2 and T1 to the other)
   b. Adjacent-2 segregation involves the movement of adjacent, homologous centromeres to the same pole (N1 and T1 to one pole, N2 and T2 to the other)
   c. Alternate segregation involves the movement of alternate centromeres to the same pole (N1 and N2 to one pole, T1 and T2 to the other)

5. In the absence of crossing-over, adjacent-1 and adjacent-2 segregation produces only duplication-carrying and deletion-carrying products, whereas all of the products of alternate segregation are balanced and therefore viable
   a. Because segregation of heterozygous translocations generally consists of 50% alternate segregation and 50% adjacent (types 1 and 2) segregation, meiosis results in 50% sterility, a condition that is called semisterility
   b. Two types of balanced, viable products occur: those with two normal chromosomes and those with two translocated chromosomes
6. Because the two translocated chromosomes must segregate together and the two normal chromosomes must segregate together in order for the meiotic products to be balanced, the genes on the two chromosomes will be genetically linked to one another

### E. Fragile X syndrome in humans

1. Fragile X syndrome, the most common form of inherited mental retardation, is associated with an abnormal X chromosome
2. Afflicted individuals generally have an X chromosome with a broken or decondensed tip
3. Not all individuals who inherit the fragile X chromosome, however, exhibit the syndrome
4. Expression of the syndrome depends on the parental origin (maternal or paternal) of the abnormal chromosome
5. This type of inheritance, in which the expression of a gene depends on its parental origin, is called ***imprinting***

## III. Changes in Chromosome Number

### A. General information

1. Chromosome mutation may involve the gain or loss of entire chromosomes or sets of chromosomes rather than chromosome breakage and rearrangement
2. Like changes in chromosome structure, changes in chromosome number also can be spontaneous or linked
3. The number of chromosomes in one set of chromosomes is called the ***monoploid*** number and is symbolized by $x$
   a. Normal ***diploids*** have two sets of chromosomes; therefore, the chromosome number of normal diploids is indicated by $2x$
   b. Because ***haploid*** cells of diploid organisms contain only one set of chromosomes, the haploid number $(n)$ equals the monoploid number $(x)$; the somatic number of chromosomes ($2n$) equals $2x$
4. Some organisms contain more than two sets of chromosomes in their somatic cells and are called ***polyploids***
   a. For these organisms, $n$ is greater than $x$
   b. For example, individuals with four sets of chromosomes in somatic cells have two sets of chromosomes in the haploid gametes; therefore, $n$ equals $2x$
   c. Individuals with three complete sets of chromosomes are called triploids
   d. Individuals with four complete sets of chromosomes are called tetraploids
   e. Individuals with five complete sets of chromosomes are called pentaploids
   f. Individuals with six complete sets of chromosomes are called hexaploids

5. Changes in chromosome number may involve one, a few, or all sets of chromosomes
6. Individuals that contain multiples of the monoploid number of chromosomes are called ***euploid***
7. Individuals that contain a change in the normal chromosome number that involves less than a full set of chromosomes are called ***aneuploid***

**B. Euploidy**

1. Euploids that contain only one set of chromosomes are called monoploids
    a. Because male ants, bees, and wasps develop from unfertilized eggs, they are monoploid
    b. In most organisms, however, monoploidy is rare
    c. Monoploid cells cannot undergo normal meiosis because of the lack of homologous chromosome pairing and subsequent random movement of chromosomes during anaphase I; the male insects mentioned above produce gametes by a modified mitotic process, thereby circumventing this limitation
    d. Monoploid plants are generated in the laboratory by inducing haploid spores to form intact plants
        (1) Such plants are valuable in plant breeding because they express recessive alleles
        (2) Monoploid plant cells can be induced to become diploid by interrupting cell division before cytokinesis
            (a) Chemical agents, such as colchicine, prevent formation of the spindle apparatus
            (b) Because the chromosomes have already replicated, interrupted mitosis yields a cell with a doubled number of chromosomes
2. Recall that polyploids contain more than two sets of chromosomes
    a. Some organisms (including flatworms, leeches, brine shrimp, some amphibians and reptiles, and many plant species) are naturally polyploid; in other organisms, polyploidy represents an aberrant chromosome number
    b. Viable polyploids that contain an odd number of chromosome sets (such as the triploids and pentaploids) generally are sterile as a result of abnormal synapsis and chromosome migration during meiosis
        (1) In a triploid, for example, two of the three homologous chromosomes may form a bivalent (a chromosome pair) while the third homolog remains as a univalent (unpaired chromosome set); alternatively, a complex trivalent may form
        (2) Anaphase I in a triploid therefore results in the production of cells with drastically altered chromosome numbers
    c. Two types of polyploids exist: *autopolyploids,* which contain multiple sets of chromosomes derived from the same species, and *allopolyploids,* which contain multiple sets of chromosomes derived from more than one species
        (1) Autopolyploids may result from spontaneous doubling of chromosomes, from the failure of chromosomes to separate during meiosis, or from chemical induction
        (2) Allopolyploidy results from the cross-hybridization of closely related species
            (a) Resulting progeny have one set of chromosomes from each parent

(b) Chromosome sets from different species are heterologous (they are not entirely homologous)

(c) If, however, the chromosome number of interspecific hybrids is doubled, the resulting cells will have two sets of chromosomes from each parent and may undergo normal synapsis

(d) In 1928, G. Karpenchenko cross-hybridized a radish and a cabbage (with subsequent chromosome doubling) to form a new hybrid species called the raphanobrassica; although the species was fertile, it did not yield the predicted cabbage-shaped head and radish-like root

d. Polyploid animals are rare; however, many plants are polyploid

(1) Modern varieties of wheat used to make bread *(Triticum aestivum)* are allohexaploid; for this plant, $2n = 6x = 42$

(2) Tetraploid flowers, vegetables, and fruits generally are larger than the corresponding diploids

(3) Triploid fruits, such as bananas and some varieties of watermelon, have the advantage of being seedless

e. In humans, triploids and tetraploids represent 15% and 5% of all spontaneous abortions, respectively

(1) Ninety-nine percent of all triploids die before birth; most of those born die during the first month of life

(2) Live tetraploid births are extremely rare

**C. Aneuploidy**

1. Aneuploidy occurs when one or a few chromosomes are gained or lost
   a. Loss of one chromosome ($2n - 1$) is called *monosomy*
   b. Loss of both homologs ($2n - 2$) is called *nullisomy*
   c. Gain of one chromosome ($2n + 1$) is called *trisomy*
   d. Gain of two chromosomes ($2n + 2$) is called *tetrasomy*
2. Aneuploidy generally is the result of nondisjunction — the failure of chromosomes or chromatids to separate during mitotic or meiotic divisions (for details about nondisjunction, see Chapter 4, The Chromosome Theory of Inheritance)
3. Aneuploidy typically is lethal in animals and deleterious in plant species
   a. Nullisomics are lethal in diploids, but may be tolerated in such polyploid plant species as wheat
   b. With the exception of sex chromosomes, monosomics also generally are lethal in animals
      (1) In humans, individuals with a single sex chromosome (X) develop Turner syndrome; the genotype of these individuals is *XO*
      (2) *XO* individuals typically are sterile females who are short in stature and have impaired cognition
   c. Several examples of trisomy are found in humans, including trisomy 13, trisomy 18, trisomy 21, and trisomy involving the sex chromosomes
      (1) Individuals with trisomy 13 (Patau syndrome) have cleft palate and lip, eye defects, congenital heart defects, and mental retardation and commonly die within the first year of life
      (2) Individuals with trisomy 18 (Edwards' syndrome) have mental retardation and an elfin appearance, including small ears, nose, mouth, and pelvis; they also generally die within the first year of life

(3) Trisomy 21 (Down syndrome) is the most common form of viable trisomy in humans
   (a) The range of phenotypes exhibited by individuals with trisomy 21 includes a lower IQ, flat face, eyelid folds, short stature, large tongue, and congenital heart defect
   (b) A small percentage of trisomy 21 results from the segregation of a translocation involving the long arm of chromosome 21, which produces duplication-carrying gametes (instead of disjunction)

(4) Three types of trisomy involving human sex chromosome may occur: *XXX* (triple X), *XXY* (Klinefelter's syndrome), and *XYY*
   (a) *XXX* individuals are females exhibiting near-normal fertility, delayed growth, and mild mental retardation
   (b) *XXY* individuals are sterile males exhibiting a tall, lanky, slightly feminized physique with slight breast development
   (c) *XYY* individuals are phenotypically normal males with few, if any, abnormal qualities

## Study Activities

1. List and describe five chromosomal landmarks that are useful for recognizing and characterizing chromosome mutation.
2. Draw a straight line to represent a chromosome. Label five loci (A, B, C, D, and E) whose distribution spans the length of the chromosome. Using this diagram to represent a wild-type chromosome, draw an example of the mutant chromosomes resulting from the following: deletion, duplication, and inversion.
3. Using the mutant chromosomes drawn in question 2, illustrate the meiotic pairing configuration that would occur in individuals heterozygous for each mutation.
4. You have at your disposal two homozygous maize lines, each containing a translocation. One line is homozygous for a translocation involving chromosomes 2 and 5, whereas the other is homozygous for a translocation involving chromosomes 2 and 10. If you cross the two lines together, what chromosomes would the resulting progeny possess? Predict how these chromosomes may pair during meiosis.
5. Explain why individuals heterozygous for an inversion or translocation commonly are associated with gametic or zygotic abortion.
6. Calculate the number of chromosomes for the following mutant cells: monoploid human, triploid human, monosomic human, and trisomic human.

# 16

# Mechanisms of DNA Repair and Recombination

## Objectives

After studying this chapter, the reader should be able to:

- Differentiate between mechanisms of direct deoxyribonucleic acid (DNA) repair and excision repair.
- Explain the significance of the SOS repair system.
- Distinguish between the breakage and reunion model of recombination and the copy-choice model of recombination.
- Describe the steps involved in the Holliday model of recombination.
- Explain the relationship between heteroduplex DNA and the process of gene conversion.

## I. Mechanisms of DNA Repair

### A. General information

1. Both prokaryotic and eukaryotic cells contain enzymes that repair damaged DNA
2. Some DNA repair systems directly repair the damaged nucleotides, whereas others simply excise the damaged area or lesion
3. Mutants that cannot repair themselves also may be deficient in terms of recombination, indicating that some enzymes involved in DNA repair also are involved in DNA recombination

### B. Direct DNA repair

1. Recall that some DNA polymerases are capable of removing mismatched nucleotides via $3' \rightarrow 5'$ exonuclease activity
2. Thymine dimers can be repaired by a ***photoreactivation*** mechanism that is catalyzed by the enzyme photolyase
3. Damage caused by alkylating agents also can be repaired directly by alkyltransferase enzymes

### C. Excision repair

1. Distortions in the DNA helix caused by thymine dimers or other lesions can be repaired via ***excision repair***
   a. A 12-base gap is created by the *Escherichia coli* multi-subunit enzyme UvrABC

b. DNA polymerase I and DNA ligase then fill the gap

2. DNA damage also can be excised by DNA glycosylase enzymes that recognize abnormal bases and remove them from the damaged nucleotide
   a. The enzyme AP (apurinic/apyrimidinic) endonuclease cleaves the sugar-phosphate backbone at the site of base loss
   b. DNA polymerase I and DNA ligase act to fill the resultant gap
3. In addition to these excision repair systems, a mismatch correction system (called mismatch repair) removes mispairing
   a. The mismatch repair enzymes mutH, mutL, and mutS identify mismatches and remove single-stranded regions that include abnormal pairing
   b. These enzymes specifically remove nucleotides from the newly synthesized strand according to the absence of base methylation

**D. The SOS repair system**

1. Many mutagens create a replication block because of the lack of specific base-pairing
2. The ***SOS bypass system,*** which is triggered when major DNA damage occurs, is the cell's emergency repair system
3. Three genes, *recA, umuC,* and *umuD,* are required for SOS repair; other genes also may be involved
4. Although the nascent DNA molecule is completed, the products from the above genes interact with DNA polymerase to alter the fidelity of replication
5. Because the bypass of a replication block frequently involves the incorporation of incorrect nucleotides, SOS repair is called error-prone repair

## II. Mechanisms of DNA Recombination

**A. General information**

1. Early models of crossing-over included the hypothesis that recombination was a result of the switching of the DNA-replication apparatus from one homolog to the other
2. This idea is called the copy-choice model of DNA recombination
3. However, the copy-choice model is inconsistent with our current understanding of recombination
   a. The copy-choice model states that only two of the four chromatids will be involved in multiple crossing-overs, whereas tetrad analysis indicates that all four chromatids may be involved
   b. Genetic and cytological evidence indicates that crossing-over occurs during prophase of meiosis I, not during the S phase of interphase
4. Evidence supporting the breakage and reunion model of chromosome recombination was presented in Chapter 7, Linkage, Recombination, and Gene Mapping
5. Recall that recombination events occur during the four-strand stage of meiosis after chromosome replication

**B. The Holliday Model**

1. The current model of recombination, called the ***Holliday model,*** was first proposed by Robin Holliday in 1964 and was modified by Matthew Meselson and Charles Radding in 1975 (for an illustration, see *Holliday Model of Recombination,* page 136)

## Holliday Model of Recombination

These illustrations depict the process of the Holliday model of recombination. Two non–sister chromatids exist (a), with pluses and minuses indicating strand polarity. One polynucleotide strand of each DNA double helix is cleaved (b), and the newly created ends form base pairs with the complementary strands of the homologous helices (c, d). Ligation of the free ends results in the formation of heteroduplex DNA (e), followed by strand migration (f). Illustrations (g), (h), and (i) show an extended form of the structure, which is rotated for clarity. Cleavage of the Holliday structure (j) may or may not result in recombination of flanking markers (k, l).

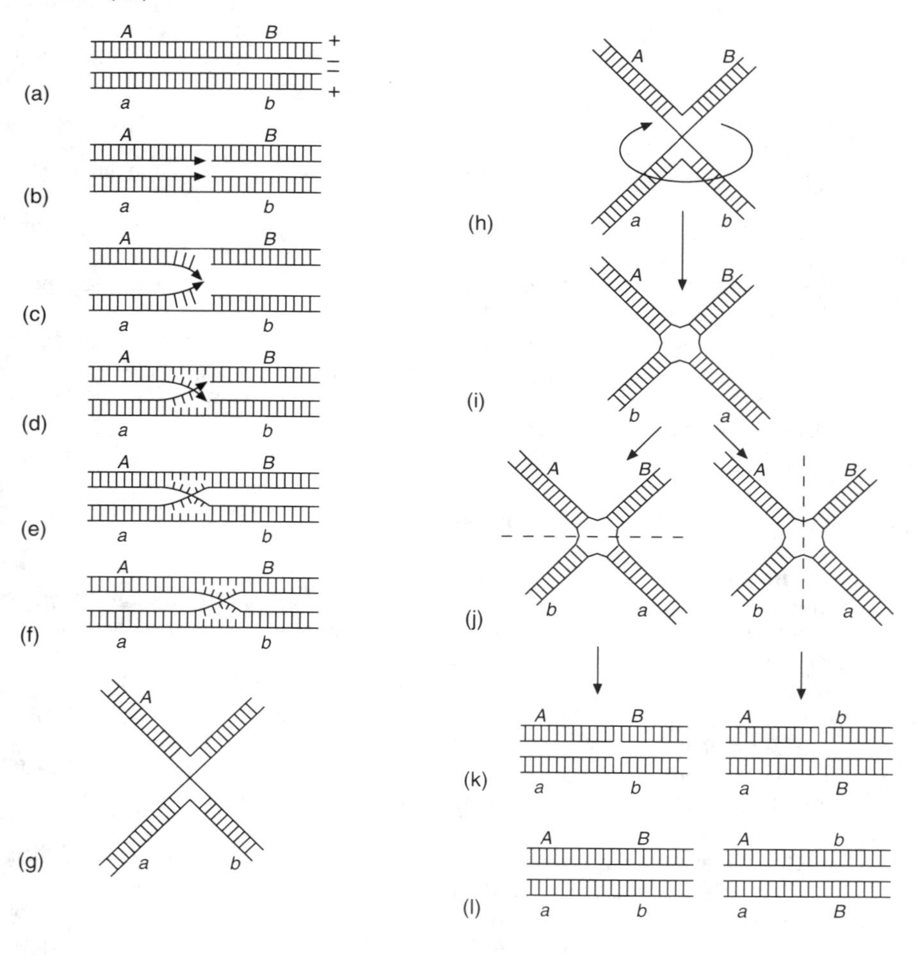


a. When two homologous double helices are aligned during synapsis, one polynucleotide strand of each DNA double helix (with the same polarity) is cleaved enzymatically
b. The newly created ends form base pairs with the complementary strands of the homologous helices; in other words, the free ends exchange pairing partners

c. Ligation of the free ends results in the formation of ***heteroduplex*** DNA, with the double helices formed by the annealing of complementary polynucleotide strands from different sources

d. This intermediate structure contains a cross bridge (or cross branch) and is called the Holliday structure

(1) A branch of the Holliday structure can move in one direction or another as hydrogen bonds are broken and reformed

(2) Branch migration increases the length of the heteroduplex region

e. Cleavage of the Holliday structure occurs in one of two ways

(1) A single cut can occur in each of the two polynucleotide strands that were originally cleaved to form the recombination intermediate

(2) A single cut can occur in each of the two polynucleotide strands that were not originally cleaved

f. After cleavage of the Holliday structure, free ends are exchanged again and strands are ligated

g. The products of crossing-over may or may not display recombination of the genes that flank the sites of cleavage and ligation

(1) If resolution involves cuts in each of the two polynucleotide strands that were originally cleaved, flanking genes are not recombined

(2) If resolution involves cuts in each of the two polynucleotide strands that were not originally cleaved, flanking genes are recombined

(3) Therefore, only 50% of initiated crossover events result in the recombination of flanking genes

2. One important aspect of the model presented above is the formation of heteroduplex DNA

a. If the homologous sister chromatids involved in the crossover event differ at the nucleotide level (that is, the organism is heterozygous), the heteroduplex region may contain nucleotide mismatches

b. Mismatch repair can result in the "conversion" of one allele to another; this process, called ***gene conversion,*** results in non-Mendelian segregation ratios

(1) One can visualize the results of gene conversion most easily in the asci of haploid fungi

(2) Recall that meiosis in a heterozygous ascomycete (*Aa,* for example) will result in two spores of one genotype *(A)* and two spores of the other genotype *(a)*

(3) If the nucleotides comprising gene *A* are included in the heteroduplex region, mismatch repair may convert a chromatid from genotype *A* to genotype *a* or vice versa

(4) The resulting ascus therefore may contain a 3:1 ratio of products rather than the typical 2:2 ratio

3. Several genes and gene products involved in recombination have been well characterized in *E. coli* experiments

a. The process of nicking and unwinding of the DNA helix is catalyzed by a protein complex whose subunits are encoded by the genes *recC, recB,* and *recD*

b. The invasion of double-stranded DNA by the single-stranded regions is accomplished with the aid of the *recA*-encoded protein

(1) Recall that the RecA protein (via the *recA* gene) also is involved in SOS repair

(2) Single-strand DNA-binding (SSB) proteins stabilize the single-stranded regions during invasion; this protein is discussed in Chapter 5, Structure and Replication of Genetic Material

c. Unwinding of the DNA and branch migration depend upon energy derived from the hydrolysis of adenosine triphosphate (ATP)

d. The RecE and RecF pathways also may be induced in some situations

## Study Activities

1. Distinguish between direct DNA repair and excision repair and give examples of both.
2. Describe the mechanism and significance of SOS repair.
3. Compare and contrast the copy-choice and breakage reunion models of recombination and indicate which is more accurate.
4. Diagram the steps involved in the Holliday model of recombination.
5. Describe the relationship between heteroduplex DNA and gene conversion.
6. List the several genes and gene products that are associated with recombination.

# 17

# Transposable Elements, Retroviruses, and Oncogenes

## Objectives

After studying this chapter, the reader should be able to:

- Describe the general characteristics of transposable genetic elements.
- Distinguish between prokaryotic insertion sequences and transposons.
- Discuss various eukaryotic transposable element families, considering each element's size, structure, and mode of transposition.
- Describe the structure and life cycle of retroviruses.
- Explain the relationship between transforming viruses, oncogenes, and proto-oncogenes.
- Give examples of proto-oncogene function.

## I. The Discovery of Transposable Genetic Elements

### A. General information

1. The basic tenets of Mendelian genetics imply that the organization of genes in a genome is invariable in the absence of mutation
2. Studies conducted during the 1950s by Barbara McClintock contradicted this view and led to the discovery of mobile genetic elements
3. In the aftermath of McClintock's groundbreaking work, geneticists have found numerous examples of mobile genetic elements in prokaryotes and eukaryotes
4. Elements that are capable of transposition (movement) within a genome are known as controlling elements, jumping genes, and ***transposable genetic elements***

### B. McClintock's controlling elements

1. McClintock's observations regarding transposable elements came from maize experiments
   a. The purple-kernel phenotype of certain maize lines is conditioned by the dominant allele *C*
   b. In heterozygous *Cc* kernels, McClintock occasionally observed patches of colorless cells that resulted from chromosome breakage and loss of the chromosome arm that carried the dominant *C* allele
   c. Chromosome breakage was caused by a factor called *Dissociation (Ds)*, which broke the chromosome at its location on the chromosome

d. However, the action of *Ds* depended upon a second factor called *Activator (Ac)*
e. Experiments designed to map *Ds* and *Ac* indicated that the genetic factors were located at different loci in different plants

2. McClintock concluded that the factors responsible for chromosome breakage, *Ds* and *Ac*, were mobile genetic elements
3. In addition to invoking chromosome breakage, *Ds* can disrupt the function of a target gene by transposing into its locus
   a. Sometimes maize kernels were rendered colorless by the insertion of *Ds* into the dominant *C* allele
   b. Unlike the colorless condition caused by chromosome breakage and loss of the dominant allele, this condition is reversible
   c. Kernels with inactivated *C* alleles had patches of colored cells that represented regions of the kernel in which the *Ds* element was excised from the *C* allele, thereby restoring gene function
   d. The transposition of *Ds* into and out of target genes depends upon the presence of *Ac*
4. Another major characteristic of McClintock's transposable elements is the ability to cause unstable mutation
   a. *Ac* was found to cause instability in the absence of *Ds*, and therefore was considered autonomous
   b. *Ds*, however, always was found to be dependent upon *Ac* for its action, and therefore is nonautonomous

## II. Prokaryotic Transposable Elements

### A. General information

1. Although transposable elements were first recognized in eukaryotes, much of our understanding of their molecular biology has come from the study of prokaryotes (bacteria)
2. Several different types of transposable elements, including ***insertion sequences, transposons,*** and certain bacteriophages (phages), occur in prokaryotic cells
3. Each type of element has different structural features

### B. Insertion sequences

1. The simplest type of transposable element that occurs in bacterial species is the insertion sequence (IS, or IS element)
2. Insertion sequences were discovered after one particular element (IS1) was found to have inserted itself into the galactose operon of *Escherichia coli*, thus creating mutant phenotypes
3. Several different IS elements have been characterized by geneticists; they are approximately 800 base pairs (bps) long and contain sequences that mobilize and integrate the element
   a. The enzyme that catalyzes transposition, called a *transposase,* is encoded by the IS element
   b. The transposase gene is flanked by terminal inverted repeats (IRs) of 9 to 41 bps that are required for transposase recognition and mobilization
4. The mechanisms by which IS elements integrate into target loci cause the duplication of a small region of the target deoxyribonucleic acid (DNA) (4 to 12 bps)

5. Two different mechanisms of transposition — conservative and replicative — exist
   a. In conservative transposition, an IS element is excised and inserted into a new locus, resulting in no net gain of IS elements
   b. In replicative transposition, an IS element is inserted into a new locus but the original element is not excised, resulting in an increase in the total number of IS elements

### C. Transposons

1. Like IS elements, transposons contain the sequences required for mobilizing and inserting an element
2. Unlike IS elements, transposons contain additional genes that have an identifiable function
3. Historically, transposons have been identified because of their ability to confer antibiotic resistance (against penicillin and tetracycline, for example)
4. The generalized structure of a transposon is more complex than that of an IS element
   a. Transposons generally contain long terminal repeats (800 to 1,400 bps) in a direct or inverted orientation
      (1) The terminal repeats actually represent intact IS elements (which are called IS modules) or sequences derived from such elements
      (2) Because the IS elements themselves contain inverted repeats, the entire transposon will contain the inverted repeats necessary for transposition
   b. The genes that confer antibiotic resistance or other characteristics are located between the IS modules
5. Transposons are found on bacterial chromosomes and on plasmids and episomes
   a. One such class of plasmids, the resistance (R) factors, are rapidly transferred during conjugation, in a manner similar to that of the *E. coli* fertility (F) factor
   b. Because R factors may contain a variety of antibiotic resistance genes, they are medically significant
6. Depending upon the IS modules contained in the element, transposons engage in conservative or replicative transposition

## III. Eukaryotic Transposable Elements

### A. General information

1. Transposable elements exist in a number of eukaryotic species, including yeast, maize, and snapdragon
2. Like the transposable elements of prokaryotes, eukaryotic elements are classified by their structure and genetic interactions with other elements

### B. The Ty elements of yeast

1. Transposable elements of yeast, called Ty elements, have variable termini and internal sequences
2. The long terminal repeats (LTRs) of Ty elements, called delta sequences, are direct rather than inverted
3. Upon integration, Ty elements cause a duplication of a small region of target DNA

4. The transposition of Ty elements involves an ribonucleic acid (RNA) intermediate, which is used as a template for the production of Ty DNA by reverse transcription
   a. The Ty element encodes reverse transcriptase
   b. Ty elements commonly are called retrotransposons because, like retroviruses, they exhibit reverse transcription

**C. The *Ac* and *Ds* elements of maize**

1. We are now in a position to understand the molecular biology of the *Ac* and *DS* elements
2. The *Ac-Ds* family is only one of many transposable element families found in maize
3. The *Ac* element is approximately 4.5 kilobase pairs (kbps) long and contains inverted terminal repeats that are 11 bps long
   a. *Ac* gives rise to a single messenger RNA (mRNA) that encodes an 807-amino acid protein
   b. Upon integration, *Ac* produces an 8-bp target site duplication
4. *Ds* elements are more variable than *Ac* elements, particularly in their interiors
   a. *Ds* elements can be considered as defective *Ac* elements; this accounts for a *Ds* element's dependence upon *Ac* for transposition
   b. Although all *Ds* elements contain the same 11-bp inverted terminal repeats, they lack the gene that encodes the transposase
5. Molecular evidence therefore confirms McClintock's initial observations
   a. Autonomous elements contain the genes required for transcriptase, along with the appropriate terminal sequences needed for mobilization
   b. Nonautonomous elements do not contain functional transposase-encoding genes, but they retain the appropriate terminal sequences for mobilization and depend on a *trans*-acting autonomous agent
6. Although the transposition of *Ac* and *Ds* does not depend upon an RNA intermediate, the transposition of this family of elements is unique because its frequency and timing are developmentally regulated

**D. Transposable elements of *Drosophila***

1. *Drosophila* contains many different types of transposable elements, three of which have been extensively characterized
2. The *copia*-like class of transposable elements consists of at least seven element families
   a. The name *copia* reflects the abundance of these elements in the *Drosophila* genome
   b. These elements are structurally similar to yeast Ty elements and range in size from 5 to 8.5 kbps
   c. Members of this class have a long, direct terminal repeat and a short, inverted repeat
   d. Like other elements examined thus far, *copia*-like elements produce target site duplication during integration
3. Another class of transposable elements, the fold-back (FB) elements, range in size from several hundred to several thousand bps
   a. Although members of this class exhibit variable nucleotide sequences, they have long, inverted terminal repeats that can fold back upon themselves

b. In some elements, the long terminal repeats represent the entire nucleotide sequence
c. FB elements may promote chromosomal rearrangements at a relatively high frequency
4. Another class of transposable element, called P, is associated with the phenomenon of ***hybrid dysgenesis,*** a syndrome observed in some *Drosophila* hybrids
   a. When females of *Drosophila* laboratory strains are mated with males from natural populations, the hybrid progeny exhibit a high rate of gene and chromosome mutation as well as sterility
   b. When males of *Drosophila* laboratory strains are mated with females from natural populations, the hybrid progeny do not exhibit hybrid dysgenesis
   c. Because many of the mutations observed in dysgenic flies were unstable, researchers hypothesized that the syndrome was associated with transposable elements
      (1) Individuals from natural *Drosophila* populations were found to contain many copies of the P element, whereas laboratory stocks did not
      (2) P elements encode both a transposase and a transposase repressor
         (a) When a low concentration of the repressor exists, transposase is produced and elements become mobile
         (b) When a high concentration of the repressor exists, transposase is not produced and elements are immobilized
      (3) Because *Drosophila* from natural populations contain P elements, repressor molecules are present in the cytoplasm of their cells (P cytotype); laboratory strains of *Drosophila* lack P elements and the repressor (M cytotype)
      (4) When an egg of cytotype M is fertilized by a sperm of cytotype P, the relative lack of repressor in the zygote leads to the mobilization of P elements and subsequent hybrid dysgenesis
      (5) When an egg of cytotype P is fertilized by a sperm of cytotype M, the relative abundance of repressor in the zygote prevents P element transposition and, therefore, dysgenesis
   d. The structure of P elements is different from that of *copia*-like and FB elements
      (1) Although full-length elements are about 2.9 kbps long, many elements appear to contain internal deletions
      (2) The elements have 31-bp inverted terminal repeats

**E. Use of transposable elements to clone specific gene sequences**

1. Once a transposable element has been cloned, it can be used as a labeled probe to isolate gene sequences into which the element has been inserted
   a. DNA from the individual in which a mutation caused by a transposable element has been identified is used to construct a genomic library
   b. A labeled, cloned element isolates clones that contain a transposable element sequence
   c. A subset of the resulting isolates will contain the gene of interest in the sequences that flank the transposable element
2. Theoretically, any gene that gives rise to a readily identifiable mutant phenotype can be cloned from organisms possessing known transposable element families

# IV. Retroviruses

## A. General information

1. Retroviruses are RNA viruses that use a DNA intermediate for replication
2. They can be considered transposable elements because they move from one location to another on the genome
3. The integrated, double-stranded virus is called a ***provirus***
4. Recall that DNA is produced from the RNA viral template by the enzyme reverse transcriptase

## B. Structure and life cycle

1. Retroviruses are structurally similar to several of the transposable element families previously discussed, such as the Ty elements of yeast and the *copia*-like elements of *Drosophila*
   a. The ends of the provirus are long terminal repeats
   b. Integration results in the production of a small target site duplication
2. Infection of a cell by a retrovirus results in the entrance of the viral capsid (which contains reverse transcriptase and viral RNA) into the host cell
   a. Once the capsid is broken down, reverse transcriptase catalyzes the formation of a DNA strand that is complementary to the viral RNA
   b. A second DNA strand is polymerized, using the first DNA strand as a template
   c. The double-stranded DNA molecule is integrated into the host chromosome
   d. For the synthesis of new viral particles, viral mRNA is transcribed from the provirus

## C. Transforming retroviruses

1. Some retroviruses can induce the growth of cancerous tumors
   a. Mouse mammary tumor virus (MMTV) and Rous sarcoma virus (RSV) are examples of such retroviruses
   b. Retroviruses that induce cancers are called transforming retroviruses or tumor retroviruses
   c. It is important to note that the term transformation is used to indicate the change to a cancer-like growth pattern in addition to describing a mode of gene transfer in bacteria
2. Viral genes that induce the growth of cancerous tumors are called viral oncogenes

# V. Oncogenes

## A. General information

1. ***Oncogenes*** are genes that cause cancer in animals
2. Viral oncogenes are found in both RNA and DNA tumor viruses
   a. For example, RSV contains the viral oncogene *v-src*
   b. Viral oncogenes are not required for viral replication
   c. Viral oncogenes most likely represent mutant cellular genes that were incorporated into the viral genome by aberrant excision from the host chromosome
3. Normal cellular genes that become oncogenes after undergoing mutation are called ***proto-oncogenes,*** or cellular oncogenes

### B. Function of proto-oncogenes

1. Although approximately 100 proto-oncogenes have been discovered already, they generally fall into one of three categories on the basis of cellular function
   a. Many proto-oncogenes encode transcription factors
   b. Other proto-oncogenes encode components of signal transduction pathways, mechanisms that alter gene activity in response to internal and external stimuli
   c. The third category of proto-oncogenes includes those that encode growth factors and growth factor receptors
2. In general, proto-oncogenes encode gene products that are important regulators of cell growth and organismal development

### C. Oncogene induction

1. Certain events convert proto-oncogenes into their oncogenic counterparts
2. Some mutations result in dominant, gain-of-function alleles; others result in recessive, loss-of-function alleles
   a. Dominant oncogenes include those that alter protein structure or increase gene activity
      (1) The *ras* proto-oncogene becomes oncogenic when a single base change converts amino acid 12 of the protein from a glycine residue to a valine
      (2) Burkitt's lymphoma has been associated with a chromosomal translocation that results in the relocation of proto-oncogene *c-myc* from chromosome 8 to chromosome 14, leading to higher rates of gene transcription
   b. Recessive oncogenes include those that eliminate proteins whose normal role is to prevent cell proliferation and tumor formation
      (1) Wilms' tumor, a cancer of the kidney, most likely is caused by the loss of a tumor-suppressor gene
      (2) Retinoblastoma, an eye cancer, also is caused by the loss of a suppressor gene
3. Therefore, oncogenes may exert their influence on cell growth by conditioning the gain of a tumor-inducing function or the loss of a tumor-suppressing function

## Study Activities

1. Explain the molecular basis for McClintock's observations of *Ac* and *Ds* behavior.
2. Make a chart that summarizes the different types of prokaryotic and eukaryotic transposable element systems; include structural features and unusual genetic behavior.
3. Compare and contrast retrotransposons and retroviruses.
4. Explain the relationship between proto-oncogenes and oncogenes.
5. Describe two mechanisms by which oncogenes can exert their effects.

# 18

## Non-Mendelian Inheritance

### Objectives

After studying this chapter, the reader should be able to:

- Distinguish between the characteristics of Mendelian inheritance and non-Mendelian inheritance and apply this knowledge to the classification of unknown inheritance patterns.
- Compare and contrast the two general types of cytoplasmic inheritance.
- Discuss the general structure of the major extranuclear genomes.
- Describe the inheritance of traits that exhibit maternal effects.
- Explain the concept of imprinting.

## I. Characteristics of Non-Mendelian Inheritance

### A. General information

1. Inheritance of some traits cannot be explained by the transmission of nuclear genes
2. Some traits demonstrate ***cytoplasmic inheritance*** as a result of the presence of extranuclear (referring to outside the nucleus) genomes
3. Other traits depend upon the genotype of the maternal parent; geneticists say that these traits show ***maternal effects***
4. Still other traits are determined by genes that are influenced by the parent from which they were derived; these genes exhibit parental ***imprinting***

### B. Identifying non-Mendelian inheritance

1. Traits that exhibit cytoplasmic inheritance, maternal effects, and imprinting do not conform to the laws of Mendelian inheritance
2. Several clues help geneticists identify non-Mendelian modes of inheritance
   a. ***Reciprocal crosses***—a pair of crosses in which the genotype of the female parent in one cross is the same as the genotype of the male parent in the other cross, and vice versa—do not yield the same phenotypes
   b. Characteristic Mendelian segregation ratios are not produced
   c. Extranuclear genes cannot be mapped to nuclear chromosomes
   d. Traits governed by extranuclear inheritance are not altered by nuclear transplantation (physical replacement of a cell's nuclear material)

## II. Cytoplasmic Inheritance and Extranuclear Genomes

### A. General information

1. The presence of two primary extranuclear genomes, the mitochondrial genome and the chloroplast genome, explains cytoplasmic inheritance
2. Mitochondria and chloroplasts are organelles responsible for adenosine triphosphate (ATP) synthesis and photosynthesis, respectively; each type of organelle has its own deoxyribonucleic acid (DNA) molecules and ribosomes
3. These two genomes are comprised of circular, double-stranded DNA molecules
4. They are present in multiple copies within the nucleoid region of the organelles

### B. Chloroplast inheritance

1. One of the first observations of cytoplasmic inheritance was that of leaf variegation in more complex plants
2. Some plants had shoots that held all green leaves, while others held leaves that were green and white (variegated) or solid white
3. Because each type of branch produces flowers, controlled crosses could be made
   a. A white-branch flower used as the female in a cross with a green-branch flower used as the male produced solid white progeny
   b. A green-branch flower used as the female in a cross with a white-branch flower used as the male produced solid green progeny
   c. If the female flower was located on a variegated branch, the progeny were variegated, regardless of the parental phenotype
4. Inheritance of leaf color was found to be strictly maternal; the paternal phenotype did not influence progeny
5. Because the green pigments of plants (chlorophylls) are located in the chloroplasts, geneticists hypothesized that chloroplast pigmentation is independent of nuclear control and that chloroplasts are inherited only from the maternal parent
6. Recall that plants with variegated leaves frequently give rise to shoots that are solid white or solid green; this indicates that chloroplasts must "sort themselves out" somehow
   a. If a zygote contains both pigmented and nonpigmented chloroplasts, ensuing mitotic divisions can be accompanied by the segregation of like chloroplasts into the same progeny cells
   b. This process, which is called cytoplasmic segregation and recombination (CSAR), essentially is mitotic segregation
7. Chloroplast genes, which are involved in CSAR, can recombine
   a. On rare occasions, the paternal contribution is not lost and a biparental zygote is produced
   b. Cells that contain organelles of different genotypes are called cytoplasmic heterozygotes or ***cytohets***
   c. By measuring the frequency of recombination, geneticists can map chloroplast genes on the circular genome

### C. The chloroplast genome

1. Each chloroplast contains multiple copies of the chloroplast DNA
2. Chloroplast DNA (cpDNA) range from 120 to 200 kilobase pairs in size

3. cpDNA contains approximately 140 genes, including ribosomal ribonucleic acid (rRNA), transfer RNA (tRNA), and protein-coding genes
   a. Some gene sequences do not have a known function, but are represented by an ***open reading frame (ORF)***
   b. An ORF is a long sequence that begins with a start codon and does not contain stop codons until the end of the reading frame
   c. The best characterized cpDNA-encoded protein is ribulose bisphosphate carboxylase (RuBP), a key enzyme in carbon dioxide fixation
      (1) RuBP represents about 50% of a green plant's protein content and therefore is the most abundant protein in the world
      (2) The functional RuBP enzyme contains eight polypeptides: four identical large polypeptides and four identical small polypeptides
      (3) The large polypeptides are encoded by chloroplast genes, whereas small polypeptides are encoded by nuclear genes
   d. The interaction of chloroplast gene products with nuclear gene products causes the organelle to function
4. cpDNA is replicated in a semiconservative manner, using chloroplast-specific enzymes
   a. The replication of cpDNA occurs throughout the cell cycle
   b. Chloroplasts do not assemble from simple components but grow and divide

**D. Mitochondrial inheritance**

1. In general, mitochondria are passed to the next generation via the female rather than the male gamete
2. Therefore, mitochondrial genes are inherited maternally
3. Some exceptions to this general rule must be noted
   a. In species exhibiting maternal inheritance of mitochondrial genomes, a slight degree of "leakage" is observed; for example, about 1 out of 1,000 mitochondria in mice are of paternal origin
   b. Some species, such as certain mussels, exhibit biparental inheritance of mitochondrial genomes
   c. Some plants, such as certain redwood species, show paternal inheritance of mitochondrial genes
4. Generally, an organism's mitochondria are genetically identical; this condition is called *homoplasmy*
5. In cases of leakage and biparental inheritance, the mitochondrial population does not have genetically identical genomes; this condition is called *heteroplasmy*
6. Mitochondrial inheritance has been studied in *Saccharomyces cerevisiae* (bakers' yeast)
   a. Like nuclear genetic analyses, mitochondrial analyses depend largely on the observation of mutant cells
      (1) Cytoplasmic petite mutants, so named because of their colony size, exhibit defects in mitochondrial electron transport
         (a) The petite phenotype is inherited in a distinct non-Mendelian fashion
         (b) Mitochondrial genomes from petite mutants are either altered in their base ratios or are completely lacking in bases, indicating that this class of mutants probably results from deletions in the mitochondrial DNA

(2) A second class of mitochondrial mutants, *mit*$^-$ mutants, also has a small-colony phenotype
(a) These mutants exhibit normal mitochondrial protein synthesis but display abnormal electron transport
(b) mit$^-$ mutants are reversible, suggesting that they result from point mutations in the mitochondrial DNA
(3) Members of a third class of mutants, ant$^R$ mutants, are resistant to antibiotics
b. When a mutant haploid cell is fused with a wild-type haploid cell, the resulting diploid cell is a cytohet
(1) As the diploid cell divides mitotically, the mitochondria sort themselves out so that after several rounds of mitosis, diploid cells contain either mutant or wild-type mitochondria
(2) The CSAR process is analogous to the one observed in chloroplasts
7. Several human diseases are caused by mitochondrial defects, but few exhibit cytoplasmic inheritance
a. All mitochondrial DNA mutations detected in humans involve genes for protein subunits of the electron transport chain
b. The resulting diseases are associated with defects of the liver, kidney, brain, and heart
c. One form of hereditary blindness, Leber's optic atrophy, is caused by a point mutation in the mitochondrial gene encoding a subunit of the enzyme NADH dehydrogenase

**E. The mitochondrial genome**
1. The amount of mitochondrial DNA (mtDNA) per genome varies widely from species to species
a. In plants, mtDNA ranges from 250,000 to 2,000,000 base pairs (bps) in size
b. Mitochondrial genomes of more complex animals generally are smaller; Drosophila contain 18,000 bps of mtDNA and humans contain 16,500 bps of mtDNA
c. However, the amount of coding sequence in a mitochondrial genome is the same for all organisms
2. In general, the function of the mitochondrial genome is to encode some of the proteins involved in mitochondrial electron transport as well as the two rRNA molecules and all tRNA molecules required for mitochondrial protein synthesis (see *Mitochondrial Genome of Humans,* page 150)
3. mtDNA specifies the production of tRNAs, rRNAs, and proteins
a. Human mtDNA encodes 2 rRNAs, 22 tRNAs, and 14 proteins
(1) The genes that encode the 12S and 16S rRNA molecules are adjacent to one another on the same strand
(2) The genes encoding the tRNAs are scattered on both strands of mtDNA; these tRNAs carry out all mitochondrial translation
(3) The protein-coding genes also are scattered on both strands
b. In less complex animals, genes may be more widely separated, with larger amounts of spacer DNA
4. In humans, only 22 tRNA molecules are required for mitochondrial translation, as opposed to at least 32 for translation of nuclear genes
a. The tRNAs encoded by mtDNA exhibit more wobble than the corresponding nuclear tRNAs

**Mitochondrial Genome of Humans**

This schematic representation of the mitochondrial genome of humans shows genes encoding transfer ribonucleic acids (tRNAs), ribosomal RNAs (rRNAs), adenosine triphosphatase (ATPase) subunits, NADH dehydrogenase subunits, cytochrome b, and cytochrome c-oxidase (CCO) subunits.

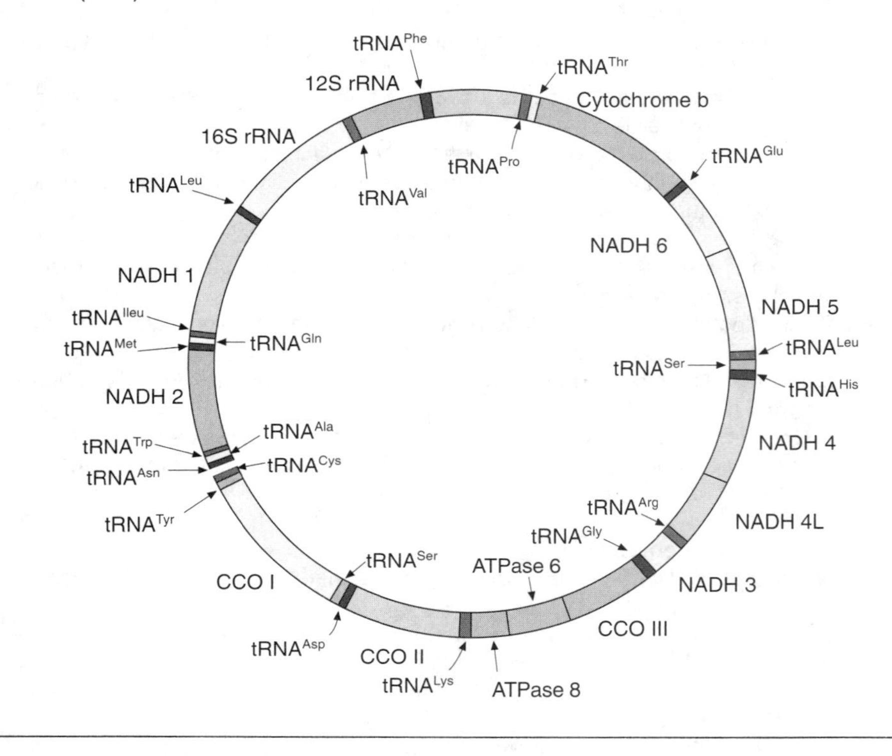


b. In addition, several codon differences exist between the nuclear genetic code and the mitochondrial genetic code as well as between mitochondrial codons of different organisms

5. mtDNA is replicated in a semiconservative manner, using mitochondria-specific enzymes
   a. mtDNA replication occurs throughout the cell cycle
   b. Like chloroplasts, mitochondria do not assemble from simple components but grow and divide

# III. Maternal Effects

## A. General information

1. A trait that exhibits maternal effects is not determined by extranuclear genes but is controlled by the expression of the maternal parent's nuclear genome
2. Maternal gene expression can lead to the formation of chemical gradients in the developing egg or embryo, thereby determining the offspring's phenotype without respect to its genotype

### B. Inheritance of maternal effect genes

1. A classic example of maternal effects involves the inheritance of snail-shell curling direction
   a. Snail shells curl either to the right (dextral coiling) or to the left (sinistral curling)
   b. Dextral coiling is conditioned by the dominant allele *(D)*; sinistral coiling is conditioned by the recessive allele *(d)*
   c. Crossing-over experiments indicate that the gene segregates in normal Mendelian ratios, but not in the expected generation
      (1) The crossing of a sinistral female with a dextral male yields all sinistral $F_1$ progeny; the crossing of a dextral female with a sinistral male yields all dextral $F_1$ progeny
      (2) $F_2$ progeny from the sinistral $F_1$ generation or the dextral $F_1$ generation are all dextral
      (3) $F_3$ progeny, however, are 25% sinistral and 75% dextral
2. Traits that exhibit maternal effects show meiotic segregation (unlike traits that exhibit cytoplasmic inheritance, which show mitotic segregation)

## IV. Imprinting

### A. General information

1. One additional mode of inheritance that does not follow the classical Mendelian rules of segregation is imprinting
2. Imprinting is the differential expression of a gene that depends upon the parent from which the alleles are derived
3. Several traits in *Drosophila,* mice, and humans (such as fragile X syndrome, Huntington's disease, and several cancers) demonstrate this unusual type of inheritance

### B. Mechanisms of imprinting

1. In imprinting, a cell recognizes from which parent its genes are derived
2. One possible mechanism in imprinting may involve differential methylation
   a. Remember that DNA bases may be modified by the addition of methyl groups (discussed in Chapter 13, Control of Gene Expression in Eukaryotes)
   b. Maternally and paternally derived gametes have different methylation patterns on their chromosomes
   c. If the level of methylation regulates the level of transcription, genes may be expressed differently depending on their source
3. A second possible mechanism in imprinting may involve differences in chromatin configuration
   a. Recall that heterochromatin is transcriptionally silent, whereas euchromatin is transcriptionally active
   b. Differential heterochromatinization may therefore regulate gene expression

## Study Activities

1. Describe how a geneticist can test for cytoplasmic inheritance.
2. Draw a pedigree that illustrates the inheritance of Leber's optic atrophy through four generations.
3. Discuss the types of genes found in mitochondrial and chloroplast genomes.
4. You have just discovered a new mutation that exhibits non-Mendelian inheritance. How can you determine whether the inheritance is the result of cytoplasmic inheritance or maternal effects?
5. You discover a snail that exhibits sinistral coiling. After self-fertilization, only dextral progeny are produced. What is the genotype of the sinistral snail?

# 19

## Quantitative Genetics

### Objectives

After studying this chapter, the reader should be able to:

- Explain the characteristics of quantitative variation.
- Discuss the concepts of polygenes and quantitative trait loci.
- Describe the role of the environment in the production of phenotypic variability.
- Use appropriate statistical methods to describe frequency distributions.
- Explain the use of correlation and regression analysis.
- Interpret estimates of broad-sense and narrow-sense heritability.
- Describe the methods used to enumerate and map quantitative trait loci.

## I. Quantitative Traits

### A. General information

1. Many heritable traits do not segregate into discrete, nonoverlapping phenotypic classes
2. Human height, birth weight, skin color, and fingerprint pattern, for example, do not divide into phenotypically distinct classes, but show continuous variation
3. Traits that exhibit continuous variation are called ***quantitative traits***
4. Quantitative inheritance is distinguished by several characteristics
    a. Traits are measured rather than counted
    b. Traits are *polygenic;* that is, they are governed by two or more genes
    c. Each ***polygene*** is located at a ***quantitative trait locus (QTL)*** and has a small, cumulative effect on a trait
    d. Traits are subject to environmental influence

### B. Polygenes

1. As stated above, polygenes influence phenotype in a limited but cumulative manner
2. To better understand the role of polygenes in quantitative inheritance, it is helpful for one to examine the effects of relatively few genes
    a. Recall that in diploid organisms, a locus with two alleles can form three genotypes (*AA, Aa,* and *aa,* for example); two loci with two alleles each can form nine genotypes *(AABB, AABb, AAbb, AaBB, AaBb, Aabb, aaBB, aaBb,* and *aabb)*
    b. The number of possible genotypes therefore is $3^n$, where n equals the number of loci

c. The number of phenotypes depends largely on the nature of the alleles and genes involved — namely, gene action
   (1) If two loci, each with two alleles, do not exhibit dominance or epistasis, the contribution of each allele is additive
   (2) For example, alleles $a^1$ and $b^1$ may increase body size by one unit above a base value, while alleles $a^2$ and $b^2$ may decrease body size by one unit below a base value
   (3) Five phenotypes will result, with organisms ranging from largest ($a^1a^1b^1b^1$) to smallest ($a^2a^2b^2b^2$) in a 1:4:6:4:1 ratio (see *The Effect of Loci Number on Frequency Distribution*)

d. As the number of loci increases, so does the number of possible phenotypic classes

e. The number of phenotypic classes equals 2n + 1, where n equals the number of loci

### C. The role of the environment

1. Recall that genotypes may exhibit various phenotypes in response to the environment
2. The range of phenotypes expressed by any one genotype in a range of environments is called the *norm of reaction*
3. If the norm of reaction for one genotype overlaps the norm of reaction for a different genotype, segregation will not result in discrete phenotypic classes
4. Therefore, the distribution of phenotypes observed for any quantitative trait will depend upon the number of polygenes, the gene action, and the effect of the environment

### D. Frequency distributions

1. A distribution of phenotypes commonly is illustrated using a frequency histogram
2. Frequency histograms based on quantitative genetic data resemble a normal distribution if the contributions of all polygenic loci are equal and additive
   a. A normal distribution resembles a typical bell-shaped distribution, with a preponderance of values near the center of the range and fewer values near the extremes
   b. Although normal distributions are symmetrical, frequency histograms based on quantitative genetic data may be symmetrical or asymmetrical
3. The single most frequent phenotypic class that occurs is called the ***mode***
4. Some frequency distributions have two modes and therefore are called *bimodal* distributions
5. Bimodal distributions may result from the mixture of two populations, each with its own mode

## II. Descriptive Statistics

### A. General information

1. The description of a frequency distribution must include information about the central tendency (the value around which individuals tend to be grouped) and dispersion (the amount of variability)
2. The mode and mean are measures of central tendency, whereas the ***variance*** and ***standard deviation*** are measures of dispersion

### The Effect of Loci Number on Frequency Distribution

Each gene is represented by two alleles with additive gene action. As the number of loci increases, the phenotypic distribution approaches a normal bell-shaped distribution.

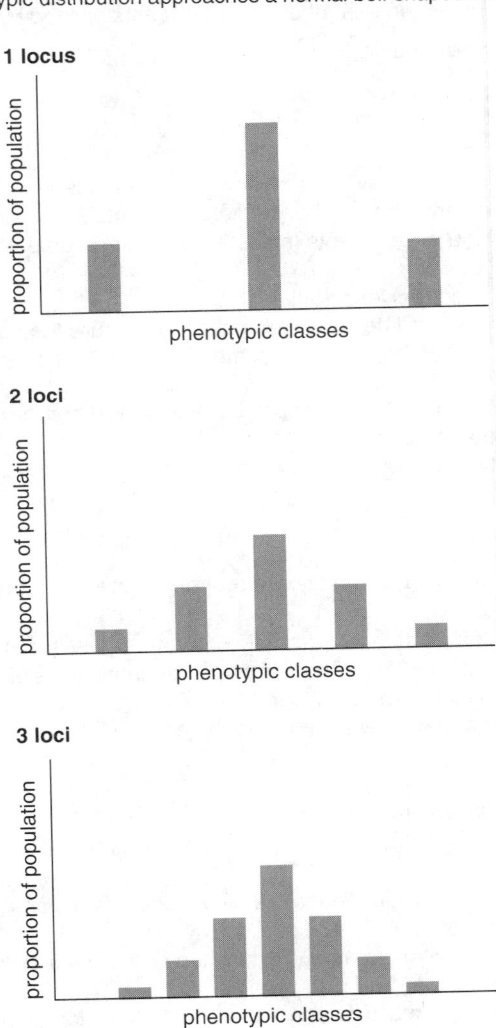


3. Geneticists rely on a subset of the population, called a sample, because it generally is impossible to measure every individual in a population
   a. Values derived from a sample of the population are called statistics
   b. Population values are called parameters
   c. A statistic is an approximation of a parameter

**B. The mean**

1. The mean (or arithmetic average) indicates the midpoint position of the distribution on the axis of measurement
2. The mean $\overline{x}$ equals the sum of all measurements divided by the number of measurements in the sample
3. Mathematically, this is written as:

$$\frac{x_1 = x_2 + \ldots + x_n}{n} \quad \text{or} \quad \frac{1}{n}\sum_{i=1}^{n} x_i$$

where n represents the number of measurements in the sample, Σ represents summation, and $x_i$ represents the individual measurement

**C. The variance and the standard deviation**

1. Variance and standard deviation are used to measure the spread of a distribution
2. Two distributions can have the same mean but different degrees of spread (for example, a wide spread or a narrow spread)
3. The variance is equal to the average squared deviation from the mean; it is symbolized as V or $s^2$
4. Mathematically, the variance is written as:

$$\frac{\sum(x_i - \overline{x})^2}{n - 1}$$

5. The standard deviation is the square root of the variance; it equals $\sqrt{V}$ or $\sqrt{(s^2)}$, or s
6. In a normal distribution, approximately 66% of the measurements are within one standard deviation of the mean, and approximately 95% of the measurements are within two standard deviations of the mean
7. Variance and standard deviation do not reveal any information about the symmetry of a distribution

**D. Correlation and regression**

1. The variability of one quantitative trait commonly is associated with the variability of another
   a. A positive relationship means that when one value increases, the other value also increases
   b. A negative relationship means that when one value increases, the other value decreases, and vice versa
2. For example, individuals with long arms tend to have long legs, whereas individuals with shorter arms tend to have shorter legs
3. In agronomic crops, the leaf mass of a plant is negatively correlated with the fruit or grain mass
4. The strength of an association — the ***correlation*** — between two traits is expressed by the correlation coefficient
   a. The correlation coefficient for two variables (x and y) is determined by first calculating the ***covariance***
      (1) The covariance equals the deviation of each observation of x from the mean of x, multiplied by the deviation of each observation of y from the mean of y

(2) Mathematically, the covariance of x and y is written as:

$$\text{Cov}(x,y) = \frac{\sum[(x_i - \bar{x})(y_i - \bar{y})]}{n - 1}$$

b. The correlation coefficient (r) is equal to the covariance divided by the standard deviations of x and y (represented by $s_x$ and $s_y$, respectively)

c. Mathematically, this is written as:

$$r = \frac{\sum[(x_i - \bar{x})(y_i - \bar{y})]}{(n - 1)(s_x\, s_y)}$$

5. The correlation coefficient will vary between –1 and +1
   a. A value of –1 represents a perfectly linear negative relationship between variables; an increase in one variable is associated with a proportional decrease in the other
   b. A value of +1 represents a perfectly linear positive relationship between variables; an increase in one variable is associated with a proportional increase in the other
   c. A value of 0 represents the absence of a linear relationship between variables
6. One must realize that correlated variables need not be identical
7. However, a correlation does not imply that a cause-and-effect relationship exists between variables
8. Geneticists rely on ***regression*** analysis to determine if a cause-and-effect relationship exists between variables
   a. Regression analysis involves the calculation of a regression line, which represents the best fit to a linear relationship between variables
   b. A regression line is defined by the equation $y = a + bx$, where a represents the y-intercept, b represents the slope of the line, and x and y represent the values of the two variables on a graph
   c. Regression analysis allows geneticists to predict the value of y for a certain value of x

## III. Heritability

### A. General information

1. Before analyzing quantitative variability, one must determine if the observed variability in a particular phenotype is influenced by genes
2. The proportion of phenotypic variation that results from genetic factors is called its *heritability*
3. Heritability estimates allow geneticists to determine the relative importance of genes and the environment for the variability of a trait
4. Variability of a trait is called the phenotypic variance and is symbolized as P or $(s_p)^2$
5. Phenotypic variance can be divided into several components
   a. The portion of the phenotypic variance that is the result of environmental (nongenetic) factors is called the environmental variance; it is symbolized as e or $(s_e)^2$
   b. The portion of the phenotypic variance that is the result of genetic factors is called the genetic variance; it is symbolized as g or $(s_g)^2$

(1) The genetic variance can be broken down into the additive genetic variance (a or $s_a^2$), the dominance variance (d or $s_d^2$), and the interaction variance (i or $s_i^2$)

(a) Additive variance is contributed by genes that exhibit additive gene action; at these loci, an allele simply adds to the effect of the other allele

(b) Dominance variance is contributed by genes that exhibit dominance; at these loci, one allele masks the expression of the other allele

(c) Interactive variance is contributed by gene interactions, such as epistasis

(2) This is written as $g = a + d + i$

c. The total phenotypic variance therefore is equal to $e + a + d + i$

**B. Broad-sense vs. narrow-sense heritability**

1. Heritability can be calculated in the broad sense and the narrow sense
2. Broad-sense heritability ($H^2$) is defined as the proportion of phenotypic variance attributed to genetic variance
   a. This is written as $H^2 = g/P$
   b. This type of heritability includes variability contributed by genes exhibiting additive gene action, dominance, and epistasis
   c. Heritability estimates range from 0 to 1
      (1) If $H^2$ equals 0, none of the observed variability in a trait can be attributed to genetic differences
      (2) If $H^2$ equals 1, all of the observed variability in a trait can be attributed to genetic differences
3. Sometimes it is more helpful to calculate heritability by using additive genetic variance rather than total genetic variance; this is called narrow-sense heritability ($h^2$)
   a. This is written as $h^2 = a/P$
   b. Because the effects of dominance and epistasis decreases the ability to accurately predict the resemblance of offspring based on the parents' phenotypes, plant and animal breeders rely on narrow-based heritability estimates when planning selection experiments
      (1) The higher the estimate for narrow-sense heritability, the higher the correlation between parent and progeny phenotypes
      (2) A high narrow-sense estimate of heritability indicates that one can shift the population mean by selecting phenotypic extremes for the breeding of subsequent generations
   c. Like broad-sense heritability, narrow-sense heritability also ranges from 0 to 1

**C. Calculating heritability**

1. Several methods are used to estimate heritability; the simplest involves the calculation of variance in genetically homogeneous populations, such as homozygous lines or crosses between homozygous lines
   a. All variation that occurs within a genetically homogeneous population is the result of nongenetic sources and therefore represents an estimate of the environmental component of the variance

b. The difference between the total phenotypic variance in genetically heterogeneous populations and the previously measured environmental variance is equal to the genetic variance; in other words, g = P – e
c. This methods yields an estimate of broad-sense heritability

2. Other heritability estimates require the comparison of phenotypes among relatives
   a. For example, we know that full siblings (individuals with both parents in common) have a 50% chance of carrying the same gene, a 25% chance that they will both have the maternal allele and a 25% chance that they will both have the paternal allele
   b. Half-siblings (individuals with one parent in common) have a 25% chance of carrying the same gene
   c. The difference in genetic correlation between full siblings and half-siblings therefore is equal to $\frac{1}{2} - \frac{1}{4}$, or $\frac{1}{4}$
   d. The difference in phenotypic correlation, however, depends on heritability
   e. The difference in phenotypic correlation between full siblings and half-siblings therefore is equal to $H^2 \times \frac{1}{4}$
   f. We can rearrange this equation such that $H^2$ is equal to the difference in phenotypic correlation between full siblings and half-siblings, multiplied by four

## IV. Quantitative Inheritance in Humans

### A. General information

1. Several human traits have received considerable attention in connection with quantitative models of inheritance
   a. Measurements of skin color in the parental, F1, and F2 generations of biracial families have led to a four-gene model for determination of skin color
   b. Other studies investigating the heritability of finger-ridge counts (fingerprints) have shown that genetic variance accounts for virtually all of the variability, with little or no environmental effect on the phenotype
2. Studies of heritability in humans occasionally are misused or misunderstood
   a. Family members frequently share a common environment as well as common genes
   b. A trait that is familial (shared by members of the same family) is not necessarily heritable
      (1) Social traits, such as religious affiliation and political party, may show a high correlation between parents and offspring but are not heritable
      (2) Linguistic differences also are not heritable even though parents and children often share the same language and dialect

### B. Heritability estimates

1. As with other studies of human inheritance, studies of quantitative inheritance are hampered by the inability to construct homozygous lines and controlled matings
2. Correlations drawn from studies involving twins are indirect methods of estimating human heritability
   a. Monozygotic (that is, from one zygote) twins who have not been brought up together provide researchers with an opportunity to address polygenic inheritance in the absence of common environments

b. Studies of twins have examined the amount of genetic variance underlying variability in such traits as IQ, susceptibility to alcoholism, and temperament

c. Generally speaking, it is extremely difficult to eliminate environmental correlations; accordingly, heritability studies in humans are subject to criticism and controversy

## V. Enumeration and Localization of Polygenes

### A. General information

1. Estimates of heritability do not provide any insight into the number or distribution of quantitative trait loci
2. Through the use of genetic markers and complex statistical tests, one can estimate the number and location of genes that contribute to the observed genetic variance

### B. Response to artificial selection

1. Artificial selection in divergent populations can lead to the establishment of homozygous lines that exhibit phenotypic extremes
   a. In 1957, James Crow selected DDT-resistant *Drosophila* by culturing a population in the presence of increasing concentrations of the insecticide
   b. Likewise, a DDT-susceptible population was maintained in the absence of the insecticide
2. Once divergent lines were produced, they were crossed to form an $F_1$ and subsequent backcross populations
   a. $F_1$ progeny were used as males in the production of backcross populations to avoid recombination of chromosomes
   b. Alternatively, crossover suppressors (such as inversions) can be used to ensure that chromosomes are transmitted to the backcross generation intact
3. In the backcross generations, individual progeny were screened for the presence of parental chromosomes via chromosome markers
   a. Correlations were sought between the presence of specific chromosomes and the appearance of parental phenotypes
   b. Crow's data indicated that each chromosome of the DDT-resistant flies could increase resistance to the insecticide, leading to the conclusion that the polygenes influencing this trait were located throughout the genome

### C. Linkage analyses

1. If divergent lines contain chromosomes that are marked with multiple genes, segregation studies can pinpoint the location of quantitative trait loci more accurately
2. The use of restriction fragment length polymorphisms (RFLPs) and other molecular markers has resulted in the production of saturated linkage maps (linkage maps with a large number of markers per chromosome)
3. By using many markers, geneticists increase the probability of detecting a linked QTL
4. These analyses are important in determining whether a trait is influenced by relatively few genes with major effects, or, conversely, many genes with relatively minor effects

## Study Activities

1. Differentiate between measures of central tendency and measures of dispersion.
2. The following numbers represent the height (in inches) of 20 students enrolled in Genetics 101 at a small college: 62, 67, 70, 64, 71, 69, 65, 69, 73, 75, 65, 68, 71, 68, 70, 69, 67, 70, 68, 69. Using this data, draw a frequency distribution and determine the mean, mode, variance, and standard deviation for this group of students.
3. Differentiate between the concepts of correlation and regression.
4. Variance components for thorax length in *Drosophila* were experimentally determined to be: $s_p^2 = 100$, $s_a^2 = 28$, $s_d^2 = 8$, $s_i^2 = 2$, $s_e^2 = 62$. Calculate the broad-sense and narrow-sense heritabilities.
5. You are a plant breeder concerned with increasing the yield of a cereal crop and have calculated the narrow-sense heritability of this trait in three populations. Estimates of $h^2$ are 0.14, 0.11, and 0.21 for populations 1, 2, and 3, respectively. In which of these populations would selection for yield be most efficient? Why?

# 20

## Population Genetics

### Objectives

After studying this chapter, the reader should be able to:
- Understand the nature and source of variation.
- Describe a population by calculating the frequencies of alleles and genotypes within the population.
- State the two major components of the Hardy-Weinberg law.
- Describe the assumptions necessary for Hardy-Weinberg equilibrium.
- Discuss the consequences of nonrandom mating.
- Explain the causes of allelic frequency changes in natural populations.

## I. Genetic Polymorphisms

### A. General information

1. All organisms contain considerable genetic variation, or polymorphism, that serves as the basis for evolutionary change
2. Population genetics is the study of this variation at the population level
3. Variation may exist within populations, between populations, or both

### B. The nature of polymorphisms

1. The relationship between genotypic and phenotypic variation differs from one trait to another
   a. Some traits, such as blood type, exhibit a simple relationship between genotype and phenotype
   b. Other traits, such as growth rate, yield, and fat content, exhibit complex relationships between genotype and phenotype
2. Experimental population genetics traditionally has utilized polymorphisms that have relatively simple relationships with the genotype
3. Such variations include morphological, immunological, chromosomal, protein, and deoxyribonucleic acid (DNA) sequence polymorphisms
   a. Morphological polymorphisms have been described throughout this book and have included differences in color, size, and shape
   b. Immunological polymorphisms include differences in antigenic specificities, such as those in the ABO blood group
   c. Chromosomal polymorphisms segregate in many plant and animal populations and include differences in chromosome morphology, the presence of

extra (supernumerary) chromosomes, and the presence of chromosome rearrangements

d. Polymorphisms at the protein level most commonly are detected after electrophoresis (and subsequent autoradiography); amino acid substitutions can result in differences in the rate of protein migration through gel

e. Variability in the occurrence of restriction sites in DNA sequences leads to restriction fragment length polymorphisms (RFLPs), which segregate in most populations examined to date

4. The ultimate source of all types of genetic variation is mutation

## C. Genotypic and allelic frequencies

1. To understand the genetic changes that underlie the evolutionary process, we must consider the genes that are shared by a group of interbreeding individuals
2. The common set of genes shared by a population at any one time is called the ***gene pool***
3. The gene pool is described in terms of the frequencies of alleles and genotypes within the population
   a. ***Genotypic frequencies*** represent the proportion of individuals of one particular genotype at one specific locus
      (1) In practice, genotypic frequency equals the number of individuals with a particular genotype in a population divided by the total number of individuals in the population
      (2) A locus with two possible alleles will give rise to three genotypes, the frequencies of which will total 1
   b. Allelic frequencies (also known as gene frequencies) represent the proportion of loci that contain a particular allele
      (1) ***Allelic frequencies*** can be calculated from the observed numbers of specific genotypes in a population and from the genotypic frequencies
         (a) When one uses the number of individuals in each genotype, the allelic frequency equals the number of copies of a given allele in a population divided by the total number of alleles
         (b) When one uses genotypic frequencies, a specific allelic frequency equals the frequency of homozygous individuals that carry the allele, plus one half the frequency of heterozygous individuals that carry the allele
      (2) For loci with only two alleles, the frequency of alleles — $f(A)$ and $f(a)$ — also are designated by the symbols p and q
      (3) For loci with only two alleles, $p + q = 1$
4. Let us consider a population segregating for alleles *A* and *a;* out of 200 individuals, 120 are *AA,* 70 are *Aa,* and 10 are *aa*
   a. The genotypic frequencies are equal to 120/200, 70/200, and 10/200, respectively; in other words, $f(AA) = 0.6$, $f(Aa) = 0.35$, and $f(aa) = 0.05$
   b. If we calculate the allelic frequencies from the observed numbers of specific genotypes, then $f(A)$, or p, equals $[(2 \times 120) + 70] \times (2 \times 200)$ or 0.775 and $f(a)$, or q, equals $[(2 \times 10) + 70] \times (2 \times 200)$ or 0.225
   c. If we calculate the allelic frequencies from the genotypic frequencies, then $f(A)$ equals $0.6 + (0.35 \div 2)$ or 0.775 and $f(a)$ equals $0.05 + (0.35 \div 2)$ or 0.225
5. The general formulas for genotypic and allelic frequencies can be expanded to accommodate populations in which more than two alleles are segregating

## II. Hardy-Weinberg Equilibrium

### A. General information

1. The ***Hardy-Weinberg law*** describes the way in which sexual reproduction influences allelic and genotypic frequencies within a population
2. The law was named after British mathematician Godfrey Hardy and German physician Wilhelm Weinberg, each of whom discovered it independently
3. The law has two major components, one that addresses allelic frequencies and another that addresses genotypic frequencies
   a. The frequencies of alleles do not change from one generation to the next
   b. After one generation of random mating, genotype frequencies will reach equilibrium
      (1) The probability that an individual will receive an *A* allele from the maternal and paternal parents is equal to $p \times p$; in other words, the frequency of *AA* equals $p^2$
      (2) The probability that an individual will receive an *A* allele from one parent and an *a* allele from the other is equal to $(p \times q) + (p \times q)$; in other words, the frequency of *Aa* equals 2pq
      (3) The probability that an individual will receive an *a* allele from the maternal and paternal parents is equal to $q \times q$; in other words, the frequency of *aa* equals $q^2$
      (4) Because the frequency of all three genotypes must total 1, $p^2 + 2pq + q^2 = 1$
      (5) In Hardy-Weinberg equilibrium, a population's genotypic frequencies remain constant from generation to generation

### B. Assumptions of the Hardy-Weinberg equilibrium

1. In order for the above relationships to be valid, four conditions, or assumptions, must be upheld
2. These assumptions are: random mating occurs within the population, the population is infinitely large, no mutation or migration exists, and no individual has a reproductive advantage over another
   a. The requirement for random mating implies that the probability that two genotypes will mate equals the product of their frequencies
      (1) Nonrandom mating may occur if individuals choose mates in a nonrandom fashion, by a process called ***assortive mating***
      (2) If mating pairs are more closely related than random pairs, ***inbreeding*** occurs; ***outbreeding*** is the mating of unrelated individuals
   b. The requirement for a large population size ensures that chance deviations from expected ratios will not occur and allelic frequencies will not be altered
   c. The requirement for the absence of mutation and migration ensures that allelic loss and addition do not occur
   d. The fact that no individual has a reproductive advantage implies that there are no evolutionary forces at work
   e. The last assumption applies only to the locus in question; other loci may be subject to evolutionary forces

### C. Extensions of the Hardy-Weinberg equilibrium

1. The calculation of allelic frequencies shown above assumed that the segregating alleles were codominant, so that all three genotypes (*AA, Aa,* and *aa*) were distinct; in other situations, fully dominant alleles are involved
   a. For situations in which one allele is fully dominant, the frequency of the dominant phenotype is equal to the frequency of homozygous dominant individuals plus the frequency of heterozygous individuals ($p^2 + 2pq$)
   b. The frequency of the recessive phenotype equals $q^2$
   c. If a population is assumed to be in Hardy-Weinberg equilibrium, the allelic frequency q can be determined by calculating the square root of $q^2$; the allelic frequency p can be determined by subtracting the value of q from 1
2. For genes with multiple alleles, the genotypic distribution can be summarized as: $p^2 + 2pq + 2pr + q^2 + 2qr + r^2$

## III. The Effects of Nonrandom Mating

### A. General information

1. Patterns of mating in natural populations are determined largely by geographical distribution and behavioral characteristics
   a. Many animals live and mate in relatively small family groups
   b. Conversely, taboos about mating of relatives reduces inbreeding in humans
2. Both inbreeding and assortive mating result in increased levels of homozygosity

### B. Inbreeding

1. If members of a mating pair are related, a gene carried by one individual may be identical to a gene carried by the other because both genes may have descended from the same DNA molecule
   a. Compared to the progeny of unrelated individuals, progeny of a related pair have an increased chance of homozygosity
   b. This phenomenon is called ***homozygosity by descent***
   c. The probability of homozygosity by descent is represented by the ***inbreeding coefficient (F)***
   d. If the individuals are more closely related, the inbreeding coefficient increases
   e. In a population with inbreeding, the proportion of *AA, Aa,* and *aa* individuals is $p^2 + Fpq$, $2pq(1 - F)$, and $q^2 + Fpq$, respectively
2. Consider the probability that an individual will be homozygous for a deleterious recessive allele that occurs at a frequency of 0.001
   a. The chance that homozygosity will result from random mating equals $q^2$, or 1 in 1,000,000
   b. The chance that homozygosity will result from full-sibling mating equals pq/4
   c. If q is 0.001, then p is nearly 1, and pq/4 is approximately equal to q/4, or 1 in 4,000
3. Close inbreeding therefore can have severe consequences
4. For example, the Amish population living in Lancaster County, Pennsylvania, is reproductively isolated from the non-Amish population because of Amish religious practices
   a. As a result of inbreeding, the Amish community exhibits a much higher frequency of Ellis-van Creveld syndrome, a recessive genetic disease

b. The frequency of this disease in the Amish community is approximately 300,000 times higher than in the general population

## IV. Changes in Allelic Frequencies

### A. General information

1. If the conditions required for Hardy-Weinberg equilibrium are not maintained, allelic frequencies, and subsequently genotypic frequencies, will be altered
2. The allelic frequencies of a population are altered by mutation, migration, random sampling error, and differential reproductive success

### B. Mutation

1. Spontaneous mutation occurs at an extremely low rate
2. Because of this low rate, mutation alone cannot account for the rapid genetic changes associated with the evolution of populations
3. Consider the mutation of allele *A*, at a frequency of $p_t$ in generation t, to allele *a*, at a frequency of $q_t$
   a. If the rate of mutation from *A* to *a* is equal to $\mu$, then the change in p that will occur in one generation is equal to $\mu p_{t-1}$
   b. This relationship tells us a that a decreasing frequency of *A* is associated with a decreased change in p per generation because fewer *A* alleles are available for mutation
   c. The ramifications of this concept become clear if we consider a concrete example
      (1) If $\mu$ equals $1 \times 10^{-5}$ (a relatively high mutation rate), the frequency of *A* equals 1, and no *a* alleles are present in the initial population, then after 10,000 generations the frequency of *a* is only 0.1
      (2) 60,000 additional generations are required for the frequency of *a* to equal the frequency of *A*
4. We conclude that mutation does not account for large changes in allelic frequency within populations even though it is the ultimate source of all genetic variation

### C. Migration

1. Movement of individuals between populations of differing allelic frequencies results in substantial changes in population structure
2. The movement of genes from one population to another is called ***gene flow***
3. The change in allelic frequency is proportional to the difference in frequency between the donor and recipient populations
4. The migration rate can be substantially greater than the mutation rate, thereby resulting in greater changes in frequency

### D. Genetic drift

1. Because all natural populations have a finite size, sampling error will prevent an allele from having exactly the same frequency in the next generation
2. The cumulative effect of sampling error over many generations is called random genetic drift or, simply, ***genetic drift***
3. Ultimately, genetic drift can result in a change of allelic frequency so that p or q equals 1
4. When p or q equals 1, no additional change in allelic frequency is possible

5. This process is especially pronounced when a small sample of individuals becomes isolated from its original population
   a. The probability that the allelic frequencies of the isolated population are exactly the same as those of the original population is very small
   b. This gives rise to a ***founder effect,*** an instance of genetic drift that can occur in a single generation
   c. Founder effect most likely is responsible for the scarcity of type B blood in Native American populations, which are comprised of descendants of small nomadic tribes

## E. Selection

1. Charles Darwin noted that organisms experience a struggle for existence in which those individuals whose phenotypes are best suited for a specific environment are more likely to survive, reach maturity, and reproduce
2. Differential survival and reproduction is called ***natural selection***
3. The relative probability of survival and the rate of reproduction is referred to as ***fitness***
   a. Fitness is a result of the interaction of a specific phenotype and a specific environment
   b. Two types of fitness, frequency-independent and frequency-dependent, exist
      (1) *Frequency-independent fitness* refers to the struggle that individuals encounter through direct interaction with the environment in the absence of competing individuals
      (2) *Frequency-dependent fitness* refers to the struggle that individuals encounter as they compete with other individuals; in this case, fitness depends upon the relative abundance of the organism
4. Alleles that are associated with higher levels of fitness will increase in a population over time
5. As the frequency of the alleles associated with higher levels of fitness increase, the average relative fitness of the population also increases
6. The effects of selection depend not only on a genotype's fitness but also on the frequencies and dominance relationships of alleles in a population
   a. For example, if the frequency of a deleterious recessive gene is relatively high, then a large proportion of homozygous individuals will express the trait and consequently will have a lower relative fitness, resulting in changes in allelic frequencies
   b. If the frequency of a deleterious recessive gene is relatively low, a small proportion of homozygous individulas will express the trait, resulting in little change in allelic frequencies
   c. The contribution of each genotype to the next generation can be calculated by multiplying the genotypic frequency and the associated fitness

## Study Activities

1. Describe the types of variation found in naturally occurring populations.
2. In a sample of 1,000 South Americans, it was found that 334, 499, and 167 individuals had blood types M, MN, and N, respectively. Calculate the allelic frequencies and the genotypic frequencies expected for Hardy-Weinberg equilibrium.
3. List the requirements for Hardy-Weinberg equilibrium and explain the ramifications of noncompliance.
4. Assume that a hypothetical population is in Hardy-Weinberg equilibrium. The number of individuals expressing the dominant allele *A* is 6,975 and the number of individuals expressing the recessive allele *a* is 3,025. Calculate the allelic and genotypic frequencies for this population.
5. The royal families of Europe experience a higher frequency of hemophilia than European families selected at random. Knowing that hemophilia is caused by an X-linked recessive allele, explain this observation.

# Appendix

# Selected References

# Index

## Appendix: Glossary

**Acrocentric chromosome**—chromosome that has a centromere nearer to one arm than the other

**Allele**—one of two or more specific forms of a gene that can exist at a locus

**Allelic frequency**—frequency at which an allele occurs within a given population

**Alternative splicing**—different way in which introns can be removed from the same mRNA molecule

**Aminoacyl-tRNA synthetase**—enzyme that catalyzes the addition of an amino acid to a tRNA molecule

**Anaphase**—phase of mitosis and meiosis in which sister chromatids or homologous chromosomes separate from each other

**Aneuploid**—condition of a cell in which there are extra or missing chromosomes; the set of chromosomes is not complete

**Antibody**—an immunoglobulin that is produced by the immune system and recognizes and binds to antigens

**Anticodon**—three-base sequence of a tRNA molecule that is complementary to an mRNA codon

**Antiparallel**—term used to describe the opposite orientation of individual polynucleotide strands in a double-stranded DNA molecule

**Assortive mating**—nonrandom mating of individuals that have similar genotypes and phenotypes; also called assortative mating

**Attenuation**—regulatory mechanism that controls the level of prokaryotic protein production through premature termination of transcription

**Autosome**—chromosome other than a sex chromosome

**Auxotroph**—an organism that requires a specific nutritional supplement for growth

**Barr body**—highly condensed and genetically inactive X chromosome present in the somatic cell nuclei of mammalian females

**Base analog**—chemical whose molecular structure is similar to a base normally found in nucleic acids

**Carrier**—an individual that possesses – but does not express – a recessive allele

**Catabolite repression**—repression of bacterial operons by a metabolic end product

**Centromere**—primary area of constriction in a eukaryotic chromosome that is involved in chromosome movement; the kinetochore is located here

**Chiasmata**—cross-shaped structures between non–sister chromatids that are visible during the latter stages of prophase I and represent the site of crossing-over

**Chromatid**—one-half of a replicated chromosome

**Chromatin**—protein that forms a complex with DNA and RNA to form a chromosome

**Chromosome**—genetic material of a cell or virus arranged in a circular or linear structure

**Cistron**—smallest unit of genetic function; synonymous with the term gene

**Clone**—identical cells arising from a common ancestor or from a segment of DNA that has been replicated after insertion into a vector molecule

**Codominance**—relationship of alleles such that the phenotypic effects of the two alleles are visible in a single heterozygous individual

**Codon**—three-base sequence of an mRNA molecule that encodes a specific amino acid or specifies the termination of translation

**Coefficient of coincidence**—ratio of the observed frequency of double crossing-over to the expected frequency of double crossing-over

**Complementary DNA (cDNA)**—DNA transcribed from an RNA molecule during reverse transcription

**Complementary gene action**—production of a dominant phenotype when different genotypes, determining similar phenotypes, are brought together in a cell or organism

**Complementation**—production of a wild-type phenotype in an individual that contains two mutant genes

**Complementation *(cis-trans)* test**—test to determine whether two mutations represent defects in the same gene or functional unit
**Conjugation**—process by which bacterial cells transfer genetic material from a donor cell to a recipient cell through direct physical contact
**Constitutive gene expression**—continuous, unregulated production of gene product (commonly a protein)
**Correlation**—degree of association between variables
**Cosmid**—vector for gene cloning composed of DNA sequences derived from both plasmids and lambda bacteriophages
**Covariance**—statistical measure of the simultaneous deviation of two variables from their respective means
**Crossing-over**—process by which corresponding segments of homologous chromosomes are exchanged through breakage and reunion
**Cytohet**—abbreviation for a cytoplasmic heterozygote, a cell that contain organelles of different genotypes
**Cytoplasmic inheritance**—inheritance governed by the extranuclear genomes of cytoplasmic organelles
**Deletion**—type of mutation that results in the loss of one or more base pairs; it also refers to a missing chromosomal segment or gene
**Deoxyribonucleic acid (DNA)**—polymer comprised of deoxyribonucleotide monomers; the genetic material of most organisms
**Dihybrid cross**—cross between two individuals that are identically heterozygous at two loci
**Diploid**—condition of a cell or organism that has two sets of chromosomes per nucleus
**DNA ligase**—enzyme that catalyzes the formation of phosphodiester bonds between adjacent $5'$-$PO_4$ and $3'$-OH groups of a DNA molecule
**DNA polymerase**—an enzyme that catalyzes the polymerization of a DNA strand in the presence of magnesium ions and a single-stranded DNA template
**Dominant**—allele that is expressed when heterozygous with a recessive allele; the phenotype of such an allele
**Duplication**—type of mutation in which more than one copy of a DNA sequence or chromosomal region exists in a chromosome or chromosome set
**Elongation factor**—protein required for the lengthening of polypeptides and translocation of ribosomes during translation
**Enhancer**—*cis*-acting regulatory DNA sequence required for maximal gene expression at a specific locus; it commonly is located several thousand base pairs away from the promoter
**Episome**—genetic elements that may replicate independently or integrate into a host chromosome
**Epistasis**—interaction in which one gene's phenotypic effects are masked by another gene at a different locus
**Euchromatin**—relatively diffuse chromosome region that has typical light-staining characteristics and can be seen only during nuclear division; it is presumed to contain active gene sequences
**Euploid**—condition of a cell or organism that contains more than one complete set of chromosomes per nucleus
**Excision repair**—process that results in the repair of DNA lesions through the removal and subsequent replacement of damaged nucleotides
**Exon**—sequence of gene's coding region that is not removed during mRNA processing and is present in the mature mRNA molecule; exons are separated by introns
**Exonuclease**—enzyme that catalyzes the removal of nucleotides from the end of a polynucleotide strand via hydrolysis of phosphodiester bonds
**Expressivity**—degree of phenotypic expression of a particular genotype
**$F_1$ generation**—generation produced by the mating of parental individuals; also called first filial generation

**$F_2$ generation**—generation produced by the mating of $F_1$ individuals; also called second filial generation
**Fertility factor (F)**—episome that facilitates the conjunction of bacterial cells by conferring the ability to donate genetic material to a recipient cell
**Fitness**—relative probability of a genotype's survival and reproductive success
**Founder effect**—change in allelic frequencies observed in a population that consists of a small, nonrepresentative sample of a larger population
**Frameshift mutation**—insertion or deletion of nucleotide pairs into a gene sequence so that mRNA nucleotides are improperly grouped into codons
**$G_1$ phase**—phase of the cell cycle that follows nuclear division and precedes chromosome replication
**$G_2$ phase**—phase of the cell cycle that follows chromosome replication and precedes nuclear division
**Gametophyte**—haploid stage of a plant's life cycle that produces gametes
**Gene**—discrete unit of hereditary information that is comprised of DNA and is located on a chromosome
**Gene conversion**—process by which one allele is converted to another allele via mismatch repair
**Gene family**—group of structurally and functionally related gene sequences of common evolutionary origin
**Gene flow**—change of allelic frequencies in a population resulting from the immigration or emigration of individuals
**Gene mapping**—determination of the linear order of loci on a chromosome
**Gene pool**—collectively, the number and variety of alleles that exist in a breeding population at a given time
**Genetic drift**—change in allelic frequencies in a population resulting from chance
**Genetics**—study of variation and heredity
**Genomic library**—collection of small, cloned DNA fragments that represents the composition of the entire genome
**Genotype**—specific allelic composition of a cell or organism
**Genotypic frequency**—frequency of a genotype at a specific locus within a given population
**Germ line**—cells whose progeny are destined to become gametes (sex cells)
**Haploid**—condition of a cell or organism that has a chromosome number that equals one-half of the normal somatic chromosome number
**Hardy-Weinberg law**—stable frequency and distribution of alleles and genotypes in a randomly mating population
**Helicase**—enzyme that disrupts the hydrogen bonds that hold together the complementary strands of a DNA double helix
**Hemizygous**—condition of a diploid cell or organism that has only one copy of a gene
**Heredity**—biological similarity between parents and their offspring
**Heritability**—proportion of phenotypic variability caused by genetic variability
**Heterochromatin**—chromosome region that is relatively condensed and of darker-staining than euchromatin; it is presumed to contain few, if any, active gene sequences
**Heteroduplex**—double-stranded DNA molecule in which the individual strands have different origins
**Heterozygous**—condition of a cell or organism that has different alleles in the sets of chromosomes
**Hfr**—bacterial cell in which a fertility factor has been integrated into the chromosome
**Histone**—class of positively charged, basic proteins that make up part of the nucleosome core particle
**Hogness box**—invariant DNA sequence with a midpoint located 25 base pairs prior to the transcriptional start site of a eukaryotic gene
**Holliday model**—proposed mechanism of reciprocal meiotic recombination
**Holoenzyme**—fully assembled enzyme that contains several subunits
**Homeotic mutation**—mutation that alters a cell's fate
**Homolog**—one member of a pair of homologous chromosomes

**Homologous chromosomes**—members of a pair of structurally and functionally equivalent chromosomes that pair during synapsis
**Homozygosity by descent**—homozygous condition in which the identical alleles in the two sets of chromosomes of a diploid nucleus probably are descended from a common DNA molecule
**Homozygous**—condition of a cell or organism that has identical alleles in the sets of chromosomes
**Hybrid**—progeny of unlike parents
**Hybrid dysgenesis**—syndrome that occurs in the progeny of crosses between natural and laboratory strains of *Drosophila* in which sterility and genetic and chromosomal mutation occur
**Imaginal disc**—group of undifferentiated cells in a blastoderm destined to become a specific adult organ
**Immunoglobulin**—proteins produced by the immune system that recognize and bind to antigens
**Imprinting**—differential expression of a gene that depends upon maternal or paternal inheritance
**Inbreeding**—mating of genetically related individuals
**Inbreeding coefficient**—probability of homozygosity by descent
**Incomplete dominance**—relationship of alleles in which the phenotype of a heterozygous individual is intermediate between the phenotypes of the corresponding homozygotes
**Independent assortment**—independent segregation of unlinked genes as a result of random alignment of bivalents during meiotic metaphase I
**Insertion sequence**—simplest type of bacterial transposable element that inserts itself in DNA
**Interference**—measure of the independence from one crossover event to another; it is equal to 1 minus the coefficient of coincidence
**Interphase**—resting phase of the cell cycle in which nuclear division does not occur
**Intragenic recombination**—crossing-over within a cistron that results in two chromosomes with different mutation sites recombining to form a wild-type gene
**Intron**—noncoding sequence of a eukaryotic gene that gives rise to an mRNA sequence that is removed during mRNA maturation
**Inversion**—mutation in which a chromosome region is removed and reinserted in the opposite direction
**Karyotype**—pictorial or photographic representation of a chromosome complement of a cell as visualized during mitotic metaphase
**Kinetochore**—region of a centromere at which spindle fibers attach during nuclear division
**Lariat structure**—lariat-shaped mRNA intermediate formed during the removal of introns
**Lethal allele**—form of a gene for which expression results in the eventual death of the organism
**Linkage**—association of genes as a result of their position on the same chromosome
**Locus**—location of a gene on a chromosome
**Lysogenic**—refers to condition of a bacterial cell that has an integrated phage
**Lytic**—refers to condition of a bacterial cell in which phage particles are reproduced and released
**Map unit**—measure of the distance between linked loci that gives rise to a 1% frequency of recombination
**Mapping function**—mathematical relationship of the distance between linked loci and the observed frequency of recombination
**Maternal effects**—influence of the maternal genotype – and not the paternal genotype – on the phenotype of the progeny
**Meiosis**—cellular process involving two successive nuclear divisions that result in the reduction of the chromosome number
**Merozygote**—bacterial cell that is partially diploid

**Messenger RNA (mRNA)**—class of RNA molecule that is translated to yield a protein

**Metacentric chromosome**—chromosome that has a centrally located centromere and two arms of equal length

**Metaphase**—phase of nuclear division in which chromosomes line up on the equatorial plane of the cell

**Missense mutation**—mutation that results in the alteration of a codon so that it encodes a different amino acid

**Mitosis**—process of nuclear division that results in the production of daughter nuclei that are identical to the parent nucleus

**Mode**—class of individuals in a frequency distribution that occurs most often

**Monohybrid cross**—cross between two individuals that are identically heterozygous at one locus

**Monoploid**—cell or organism that has one complete set of chromosomes

**Mosaicism**—condition in which a tissue or organism contains two or more genetically distinct cell types

**Mutagen**—agent that induces mutation

**Mutant**—adjective describing a gene, cell, phenotype, or organism other than the wild-type

**Mutation**—process of change in the genetic material; the gene or chromosome that results from such a process

**Natural selection**—process in which individuals of one genotype contribute more progeny to the next generation than individuals of other genotypes

**Nondisjunction**—cellular process in which there is a lack of separation of homologous chromosomes or sister chromatids during the anaphase portion of nuclear division

**Nonparental ditype (NPD)**—class of tetrad that contains two types of meiotic products, both of which represent recombinant genotypes

**Nonsense mutation**—mutation resulting in the alteration of a codon that signals termination of translation rather than encoding of an amino acid

**Norm of reaction**—range of phenotypes expressed by a specific genotype in different environmental conditions

**Northern blotting**—technique in which electrophoretically separated RNA is transferred onto a medium to prepare for hybridization with a labeled probe

**Nucleolus organizer region (NOR)**—chromosome segment that is physically associated with the nucleolus, which contains the genes encoding rRNA

**Nucleoside**—molecule containing a nitrogenous base and a sugar group

**Nucleosome**—complex of eight histone molecules wrapped by approximately 146 base pairs of DNA

**Nucleotide**—molecule containing a nitrogenous base, a sugar group, and a phosphate group; unit of structure in nucleic acids

**Okazaki fragments**—relatively short segments of newly replicated DNA that result from discontinuous DNA replication

**Oncogene**—gene that causes cancer

**Open reading frame (ORF)**—sequence of DNA flanked by an initiation codon and a termination codon that presumably represents the coding sequence of a gene

**Operator**—region of an operon at which trans-acting regulatory proteins bind; it determines whether structural genes are transcribed

**Operon**—cluster of coordinately regulated, adjacent genes that are transcribed as a single mRNA molecule

**Outbreeding**—mating of genetically unrelated individuals

**Paracentric inversion**—inversion mutation in a chromosome regions that involves only one arm and not the centromere

**Parental ditype (PD)**—class of tetrad that contains two types of meiotic products, both of which represent nonrecombinant (parental) genotypes

**Parental (P) generation**—generation containing individuals from pure lines to be used in producing an $F_1$ hybrid

**Pedigree**—family-tree diagram that indicates inheritance of specific phenotypes

**Penetrance**—proportion of individuals with a specific genotype that express the corresponding phenotype
**Peptidyl transferase**—enzyme that catalyzes the formation of peptide bonds during translation
**Pericentric inversion**—inversion mutation in a chromosome region that involves both arms and the centromere
**Phenotype**—physical manifestation of a genotype in a particular environment
**Photoreactivation**—light-dependent enzymatic mechanism by which thymine dimers are repaired
**Plaque**—clear area in a bacterial lawn created through lysis of bacterial cells after bacteriophage infection
**Plasmid**—autonomously replicating extrachromosomal genetic element
**Pleiotropy**—phenomenon in which several apparently unrelated phenotypes are associated with a single mutation
**Point mutation**—mutation caused by the substitution of one base for another in a base pair
**Polygene**—one of a large number of genes that has a small, additive effect on the phenotype of a quantitative trait
**Polymerase chain reaction (PCR)**—technique for amplifying specific DNA sequences that involves repeated cycles of denaturation, annealing of primers, and primer extension
**Polyploid**—cell or organism that has three or more complete sets of chromosomes
**Polytene chromosome**—large chromosomes consisting of many chromatids that are held together after many rounds of chromosome replication in the absence of nuclear division
**Precursor RNA (pre-RNA)**—primary transcript of a gene that will subsequently be processed
**Pribnow box**—invariant DNA sequence with a midpoint located 10 base pairs prior to the transcriptional start site of a prokaryotic gene
**Primase**—enzyme responsible for catalyzing the synthesis of an RNA primer for use in the initiation of DNA synthesis
**Probability**—expected frequency of a particular event; it is calculated as the number of times an event is expected to happen divided by the number of opportunities for that event to occur
**Product rule**—probability that two independent events will occur simultaneously is equal to the product of the individual probabilities
**Promoter**—*cis*-acting regulatory region of a gene at which RNA polymerase binds to DNA to initiate transcription
**Prophage**—bacteriophage DNA that has been integrated into the host chromosome
**Prophase**—phase of nuclear division during which chromosomes condense
**Proto-oncogene**—gene that, when altered, becomes an oncogene
**Prototroph**—organism that does not require a specific nutritional supplement for growth
**Provirus**—viral DNA that has been integrated into the host chromosome
**Pseudogene**—formerly active that has been rendered inactive (nonfunctional) by mutation
**Punnett square**—diagrammatic representation of allele combinations resulting from a cross of two individuals
**Purine**—type of nitrogenous base that contains a double-ringed base; represented in DNA by adenine and guanine
**Pyrimidine**—type of nitrogenous base that contains a single-ringed base; represented in DNA by cytosine and thymine and in RNA by cytosine and uracil
**Quantitative trait**—trait that exhibits continuous variation
**Quantitative trait locus (QTL)**—locus at which genetic variability is correlated with the phenotypic variability of a quantitative trait
**Reannealing**—re-forming of hydrogen bonds between complementary polynucleotide strands after denaturation

**Recessive**—allele that is not phenotypically expressed in a heterozygote, or the phenotype of such an allele
**Reciprocal crosses**—pair of crosses in which the genotype of the female parent in one cross is the same as the genotype of the male parent in the other, and vice versa
**Recombinant**—refers to a chromosome, cell, or organism exhibiting a nonparental combination of genes as a result of recombination
**Recombination**—process that generates nonparental combinations of genes via crossing-over or independent assortment
**Regression**—statistical measure of the linear relationship between variables
**Regulatory gene**—gene encoding a protein product that is involved in the activation or repression of structural gene transcription
**Replica plating**—procedure for transferring bacterial colonies from a master plate to a new plate while maintaining the relative positions of the colonies
**Repressor protein**—protein that prevents transcription of an operon by binding to the operator region
**Restriction endonuclease**—enzyme that recognizes specific DNA sequences and subsequently cleaves the molecule
**Restriction mapping**—determination of the recognition sites of specific restriction endonucleases within a DNA segment
**Restriction site**—site at which a specific restriction endonuclease breaks a DNA molecule
**Retrovirus**—RNA virus that replicates via a DNA intermediate; in other words, virus that directs RNA-dependent DNA synthesis
**Reverse transcriptase**—enzyme that catalyzes the polymerization of a DNA molecule from an RNA template
**Rho**—protein required for termination of the transcription of certain genes in *Escherichia coli*
**Ribonucleic acid (RNA)**—polymer comprised of ribonucleotide monomers that generally are single-stranded; the genetic material of some viruses
**Ribosomal RNA (rRNA)**—RNA contained in ribosomes
**Ribozyme**—RNA molecule that exhibits enzymatic activity
**RNA polymerase**—enzyme that catalyzes the polymerization of an RNA strand from a DNA template and functions in transcription
**RNA replicase**—enzyme that catalyzes the polymerization of an RNA strand from an RNA template
**RNase**—enzyme that catalyzes the hydrolysis of RNA molecules' phosphodiester bonds
**S phase**—phase of the cell cycle in which DNA is synthesized and chromosomes are replicated
**Segregation**—separation of homologous chromosomes and alleles into different cells during meiosis
**Semiconservative replication**—process of DNA replication in which each daughter DNA molecule contains one parental strand and one newly synthesized strand
**Sex chromosome**—chromosome that is correlated with the sex of an individual
**Sexduction**—infectious transmission of bacterial genes via the fertility factor
**Sex linkage**—inheritance pattern exhibited by loci on a sex chromosome
**Sigma factor**—protein that provides RNA polymerase with the ability to recognize promoters
**Silencer**—*cis*-acting regulatory element that has properties similar to those of an enhancer, but represses transcription of eukaryotic genes
**Small nuclear RNA (snRNA)**—class of eukaryotic RNA molecule that functions in the processing of pre-RNA molecules
**SOS bypass system**—error-prone process of DNA repair in which imprecise polymerization permits the bypass of damaged regions
**Southern blotting**—technique in which electrophoretically separated DNA is transferred onto a medium in preparation for hybridization with a labeled probe

**Splicesome**—protein-RNA complex that removes introns from eukaryotic mRNA molecules
**Sporophyte**—diploid stage of a plant's life cycle in which haploid spores are produced via meiosis
**Standard deviation**—measure of dispersion calculated as the square root of the variance
**Stem-loop structure**—structure formed by the folding back of a single-stranded nucleic acid upon itself so that a double-stranded stem is capped by a single-stranded loop
**Structural gene**—gene that encodes enzymes
**Submetacentric chromosome**—chromosome with a centromere slightly nearer to one arm than the other
**Sum rule**—probability that one of several mutually exclusive events is the sum of the individual probabilities
**Supercoiling**—twisting of a double-stranded DNA molecule on itself
**Suppressor mutation**—mutation that counteracts the expression of a mutation at a particular site
**Synapsis**—pairing of homologous chromosomes during prophase I of meiosis
**Syncytium**—presence of many nuclei in one cell's cytoplasm
**Telocentric chromosome**—chromosome with a centromere at one end
**Telomere**—end of a chromosome that has a specific DNA sequence
**Telophase**—phase of nuclear division during which daughter nuclei typically are formed
**Testcross**—experimental cross of an unknown genotype with an individual that is homozygous recessive at the locus (or loci) of interest
**Tetrad**—association of four homologous chromatids that occurs during prophase of meiosis; the four products of a single meiotic event
**Tetrad analysis**—genetic analysis of the four products of a single meiotic event
**Tetratype (T)**—class of tetrad that contains four types of meiotic products, with two types representing recombinant genotypes and the other two types representing parental genotypes
**Topoisomerase**—enzyme that catalyzes the supercoiling of DNA double helices
**Transcription**—process by which RNA is synthesized from a DNA template
**Transcription factor**—protein that directly or indirectly regulates the initiation of transcription by binding to *cis*-acting regulatory sequences of a gene
**Transduction**—process of DNA transfer from one cell to another through viral infection
**Transfer RNA (tRNA)**—RNA molecule that brings amino acids to the site of protein synthesis
**Transformation**—process of DNA transfer in which recipient cells import extracellular, exogenous DNA fragments, resulting in a different genotype; or the change from a normal eukaryotic cell to a cancerous one
**Transition mutation**—point mutation in which a purine is substituted for another purine or a pyrimidine for another pyrimidine
**Translation**—process in which a protein is synthesized
**Translocation**—chromosome mutation in which two different chromosomes are broken and reunited
**Transposable genetic element**—sequence of DNA that can be excised and reinserted into a different locus
**Transposon**—see *Transposable genetic element*
**Transversion mutation**—point mutation in which a purine is substituted for a pyrimidine or a pyrimidine for a purine
**Variance**—measure of dispersion that is equal to the average squared deviation from the mean
**Variation**—biological differences between parents and their offspring or between individuals of a population
**Wild-type**—refers to genotype or phenotype found in nature; the "normal"genotype or phenotype

## Selected References

Cummings, M.R. *Human Heredity: Principles and Issues* (3rd ed.). St. Paul: West Publishing Co., 1994.

Gardner, E.J., Simmons, M.J., and Snustad, D.P. *Principles of Genetics* (8th ed.). New York: John Wiley & Sons, 1991.

Griffiths, A.J.F., Miller, J.H., Suzuki, D.T., Lewontin, R.C., and Gelbart W.M. *An Introduction to Genetic Analysis* (5th ed.). New York: W.H. Freeman and Co., 1993.

Klug, W.S., and Cummings, M.R. *Concepts of Genetics* (4th ed.). New York: Macmillan, 1994.

Lewin, B. *Genes IV.* Oxford, England: Oxford University Press, 1994.

Russell, P.J. *Genetics* (3rd ed.). New York: Harper Collins Publishers, 1992.

Singer, M., and Berg, P. *Genes and Genomes.* Mill Valley, Calif.: University Science Books, 1991.

Tamarin, R.H. *Principles of Genetics* (4th ed.). Dubuque, Iowa: William C. Brown Publishers, 1993.

Watson, J.D., Gilman, M., Witkowski, J., and Zoller, M. *Recombinant DNA* (2nd ed.). New York: W.H. Freeman and Co., 1992.

Weaver, R.F., and Hedrick, P.W. *Basic Genetics: A Contemporary Perspective.* Dubuque, Iowa: William C. Brown Publishers, 1991.

Woodward, V. *Human Heredity and Society.* St. Paul: West Publishing Co., 1992.

# Index



i refers to an illustration; t, to a table

i refers to an illustration; t, to a table

i refers to an illustration; t, to a table

## Notes

# Notes